JUN 1 3 2003

COMMUNITY DEVELOPMENT AND SCHOOL REFORM

ADVANCES IN RESEARCH AND THEORIES OF SCHOOL MANAGEMENT AND EDUCATIONAL POLICY

Series Editor: Gary M. Crow

ADVANCES IN RESEARCH AND THEORIES OF SCHOOL
MANAGEMENT AND EDUCATIONAL POLICY   VOLUME 5

COMMUNITY DEVELOPMENT AND SCHOOL REFORM

EDITED BY

ROBERT L. CROWSON

Peabody College, Vanderbilt University in Nashville, USA

2001

JAI
An Imprint of Elsevier Science

Amsterdam – London – New York – Oxford – Paris – Shannon – Tokyo

ELSEVIER SCIENCE Ltd
The Boulevard, Langford Lane
Kidlington, Oxford OX5 1GB, UK

First edition 2001

        Library of Congress Cataloging in Publication Data

Community development and school reform / edited by Robert L. Crowson.
        p. cm. – (Advances in research and theories of school management and educational policy; v. 5)
        Includes bibliographical references.
        ISBN 0-7623-0779-X
        1. Community and school. 2. Educational change. I. Crowson, Robert L. II. Series.

        LC215 .C5585 2001
        371.19–dc21   2001029750

British Library Cataloguing in Publication Data
A catalogue record from the British Library has been applied for.

ISBN: 0-7623-0779-X

⊚ The paper used in this publication meets the requirements of ANSI/NISO Z39.48-1992 (Permanence of Paper). Printed in The Netherlands.

CONTENTS

LIST OF CONTRIBUTORS

Robert L. Crowson — Department of Leadership and Organizations, Vanderbilt University

Mary Erina Driscoll — School of Education, New York University

Ellen B. Goldring — Department of Leadership and Organizations, Vanderbilt University

Charles Hausman — Department of Leadership and Policy, University of Utah

Charles Taylor Kerchner — Claremont Graduate University

Hanne B. Mawhinney — College of Education, University of Maryland

Charis L. McGaughy — Department of Leadership and Organizations, Vanderbilt University

Cristall McGill — Division of Leadership and Policy Studies, Arizona State University

Grant McMurran — Claremont Graduate University

Louis F. Miron — Department of Education University of California at Irvine

Robert A. Peña — Division of Leadership and Policy Studies, Arizona State University

Dennis Shirley   *Lynch School of Education*
                           Boston College

Claire Smrekar   *Department of Leadership and*
                           Organizations, Vanderbilt University

Robert T. Stout   *Division of Leadership and*
                           Policy Studies, Arizona State University

Rodolfo D. Torres   *Department of Education*
                           University of California at Irvine

COMMUNITY DEVELOPMENT AND SCHOOL REFORM: AN OVERVIEW

Robert L. Crowson

INTRODUCTION

In his discussion of late nineteenth-century rural education in the midwest, historian Wayne Fuller (1982: 73) wrote:

> ... the country schoolhouse on the Middle Border was no architect's dream, but it was a monument to community enterprise, or lack of it, and its appearance was one measure of the district's interest in education.

Though always "functional," the country school was often impressively all-brick, clean, and substantial. But, it was even more likely to be poorly constructed, uncomfortable, dirty, and covered by unpainted clapboard of the lowest quality. Nevertheless, the schoolhouse and its grounds, tended to improve in attractiveness, observed Fuller (1982: 76), 'as the farmers' fortunes improved.'

Interestingly, after many decades of a different set of foci in public education, the reciprocal link between schooling and each neighborhood's "fortunes" – is now undergoing careful re-examination. As Driscoll and Kerchner (1999) put it, public schooling is busily rediscovering its "sense of place" – not just as a provider of instructional services to a community but additionally as a key organization in the most basic economy of each neighborhood. "Understanding neighborhoods as zones of production," they write, is just as important to educational administration as the developmental, pedagogical, and social-service

Community Development and School Reform, Volume 5, pages 1–18.
ISBN: 0-7623-0779-X

theorizing that long has described the institution of the individual school (Driscoll & Kerchner, 1999: 396).

It is a far cry, however, from realizing anew that the fortunes of neighborhood and public school are fundamentally intertwined – to a full understanding of just how "production" within both schoolhouse and community can be encouraged and shared. As its major purpose, this book seeks to explore, in some depth, an emerging set of interests in the potentiality of community-level social and economic development – as an important influence upon school reform; and vice versa.

Indeed, in her pathbreaking work on this topic, Lizbeth Schorr (1997: 291) has cleanly and succinctly identified the central thrust of our own endeavor – in noting that a realization increasingly is that educational success necessitates a key place for the school ". . . at the table where community reform is being organized." But, just what it takes for the schools to sit at such a table, and for the table to become rich and bountiful as well as educationally wholesome – remains, at this time, much a mystery.

THE "PROBLEM"

The rediscovery of the power of community-development connections has contemporary roots in both "services" and decentralization movements in public education. To be sure, there are century-long antecedents to coordinated children's services experimentation; and a community-control approach to school district decentralization received considerable, and often heated attention in the mid-1960s. Nevertheless, the spread of coordinated-services initiatives and site-level empowerment efforts has in both cases been fairly recent – owing much to such influential people, organizations, and events as: James Coleman, Sharon Kagan, James Comer, Michael Kirst, Lizbeth Schorr, Joyce Epstein, the Annie E. Casey Foundation, Chicago School Reform and Kentucky School Reform (see Crowson & Boyd, 1993).

The services movement has achieved a steady growth and increased acceptability among educators. The shorthand descriptor FRC, standing for Family Resource Center is now commonly recognized. FRC's and similarly identified entities are to be found in areas of family need throughout the nation, in both rural and urban settings. The services movement, however, has not met many of its public and professional expectations to date. Much rooted in the powerful notion of social capital development, direct linkages between added services to children and families and improved education or educational opportunities for children have yet to be firmly established (see White & Wehlage, 1995; Smrekar & Mawhinney, 1999). The services approach has perhaps generated greater acceptability and legitimacy than effectiveness, thus far. Thus, there are questions about the movement's long term sustainability.

As an element in the analysis, there is now a developing recognition, as particularly noted by Schorr (1997), that service integration by itself may fall well below the full scope of efforts needed to improve learning opportunities in low-income environments. Much more than just formal, specialized services is required, argues Schorr (1997). Opportunity:

> . . . also depends on the creation of informal, helping networks, including church and social ties, family support services, youth development programs, mentoring, recreational opportunities, and strong bonds among adults (Schorr, 1997: 287).

Additionally, studies of services initiatives to date have identified an array of often-unresolved professional, managerial, and organizational issues as barriers to success. Dilemmas of professional turf, resistance to change, cross-agency collaboration and socio-cultural gaps between service-providers and service-recipients – are among the key constraints unearthed thus far (see Crowson & Boyd, 1993; White & Wehlage, 1995; Smrekar & Mawhinney, 1999).

Without denying the power and the importance of the services thrust, the community-development notion draws much of its appeal from an altered conception of the central "problem" of school improvement. To begin with, the provision of outreach through "services" to members of a neighborhood limits an appreciation of the back-and-forth strengths between school and community. Community development theorizing suggests that services are a two-way street – at least as much community-to-school as they are school-to-community. The point is thoroughly exemplified in today's rapidly growing literature on the value of out-of-school time services and opportunities (as these are provided by libraries, Boys' and Girls' clubs, churches, drop-in centers, etc.) (see Behrman, 1999). The many institutions of the community, offering an array of activities far beyond those typically identified as "school," can all play an impor-tant, interactive role in child care and development (see Vandell & Shumow, 1999).

Because so many services initiatives are one-way, warned White and Wehlage (1995), the result frequently is a set of professionally identified "needs" and a *supply* of assistance to families – but the neglect of family-identified needs and the neighborhood's *demand* for services. A key, yet-unmet test for community services, continued White and Wehlage (1995) is less their success in garnering and delivering services than it is in engendering a closer understanding of, and even a partnership with, the clientele to be served.

A second altered conception of the problem views "service" as a severely limiting agenda in communities that frequently are seeking broadened leadership and revitalization. A services agenda may be necessary but not sufficient – especially in neighborhood circumstances that also lack resources in the developmentally important arenas of recreation, housing quality, employment opportunity, law enforcement, etc. (see Littell & Wynn, 1989; Haveman & Wolfe, 1994). Beyond services, the larger agenda (into "development") typically includes much discussion of self-help, empowerment, enterprising, investment, and public-private cooperation (see Garr, 1995). Such an agenda creates spin-off problems of its own, of course – for professional service-providers are traditionally much more comfortable with a language of "meeting needs" above the emerging language of market-forces, enterprising, regeneration, and entre-preneuralism.

Finally, a third redefinition of the problem asks anew just what it means to engage in *school* improvement. In addition to an outreach with family services, many educators across the nation are vigorously involved in building bridges across the boundaries between profession and clientele. Parent involvement and more effective school-community relations are 'hot' topics currently, within both the practitioner and the academic literatures (see Smrekar & Mawhinney, 1999). Old, tried-and-true administrative practices to "buffer" the schools from intrusion are losing credibility amid a growing recognition that "open" schools can benefit enormously from parental input.

While there are school effects on families, to be sure, an appreciation and understanding of family effects on the schools is just beginning to receive sophisticated scholarly attention (see Booth & Dunn, 1996). Recent research by Leithwood and Jantzi (1999), for example, has discovered that the within-school sources of leadership in student "engagement" with school are more likely to reside in family educational culture than with teacher or prin-cipal influences. Our understanding of school leadership, they concluded is:

> ... unlikely to progress much further without systematic inquiry about how schools and families coproduce the full array of outcomes for which schools are responsible (Leithwood & Jantzi, 1999: 699).

BACKGROUND

The rapidly increasing interest in community-development activity may help to redefine the central problem of school improvement into a more ecological conception of "coproduction" between school and community. However, there is much yet to be learned about both the practice and the developmental potential of community-revitalization strategies – and even more to be learned about connections with the schools. A set of thorough understandings are far from clear – as to just what it might mean to move toward a community-based foundation for school-improvement and toward revitalizing linkages between schools and their surrounding neighborhoods.

"Going local" is by no means a new idea, observes Michael Shuman (1998). Perhaps it appears rather strange today, at a time of globalizing economies and merger mania. On the other hand, it may not be strange at all to re-investigate ideas of community self-reliance amid a set of expanding economic forces. Interestingly, turn-of-the-century neighborhood initiatives (e.g. the settlement house movement) occurred in tandem with progressive reformers' efforts to centralize and standardize city governments (Halpern, 1995). And, currently, a decided renewal of interest in neighborhood schooling is to be found alongside an accelerating state and federal press toward subject-matter standards and high-stakes' accountability testing (Crowson, 1999).

"Going local" today is becoming increasingly complex and most intriguing – complete with a rich set of opportunities for deepened theorizing around the ecology of school-community connections. Indeed, there may be more than just a little nostalgia for the old time schoolhouse and a restored 'sense' of community wrapped around this movement.

However, the modern-day version of going local is also replete with some emergent (and potentially fruitful) conceptual lenses into the school and community relationship. Among the constructs with high potential are:

(1) a re-examination of "place" in school improvement;
(2) a deepened inquiry into a community's array of "institutional investments" in school reform;
(3) an added understanding of community "empowerment" as a force in school improvement (particularly within the framework of that which is described as "urban regime theory" [see Elkin, 1987; Ramsay, 1996]); and,
(4) further knowledge about coproductive "partnering" between school and community (within the context of community-development).

Place

"Location, location, location" claims the real estate industry. "There's no place like home," goes the old refrain. "Our deepest yearnings," notes Shuman (1998: 31), 'are linked to a sense of place.' Places have a special 'preciousness' for us, add Logan and Molotch (1987: 17).

Historically, the schoolhouse was one of the few social institutions rural people encountered daily, concluded David Tyack in *The One Best System* (1974: 17); and the schoolhouse "both reflected and shaped a sense of community." Moreover, a very common cause for argument in rural America, over the years, "was the location of the school" (Tyack, 1974: 17). We sometimes fail to remember that schools are "embedded" in their communities, observe Driscoll and Kerchner (1999).

Place matters. Place is far more than geography. It is a source of personal values; of sentiments, of growth and hope; of spirituality; of interactions and mutual support; of familiarity, security, and trust; of shared symbols and special ties; of reputation and identity (Shuman, 1998). Place can also, of course, be a source of frustration, limited horizons and little support, insecurity and even danger, and a deep urge to escape. Places can also be thought of as commodities, as an assemblage of needs, as healthy or unhealthy, as resourceful or barren, as integrated or disintegrated, as stable or in transition.

Nevertheless, despite all of these associations, it is not entirely clear just how place matters. In suggesting a "new economics of place," Shuman (1998: 45–50) urges a drive toward self-reliance at the local level – a minimizing of dependence through local ownership and employment, neighborhood banking, resources reinvested in the local community, "real home rule," community corporations, and "inducing residents to prefer local goods and services." Driscoll and Kerchner (1999: 398) come fairly close to much the same perspective, in suggesting 'schools as engines' of a developing community economy – hiring and purchasing locally and "connecting school resources to community development" (see also Kerchner, Koppich & Weeres, 1997).

Beyond an economics of place, there has not been a great deal of attention to date to a pedagogy of place. There is, to be sure, a solid sense of place in much of the thinking around creating and sustaining social capital – from James Coleman (1988) forward. This construct has been a powerful tool in conceptualizing school outreach with parent assistance and engagement, family partnerships, coordinated children's services, and the like (Driscoll & Kerchner, 1999).

There has been some important work in recent years, in "unpacking" the construct of social capital development into key elements in a strategy of community support – elements such as *access to information* (e.g. about jobs, programs for youth, family services, etc.); *direct services* (e.g. tutoring, counseling, recreation); *norms* (e.g. academic expectations, pro-social behavior); *social trust* (e.g. relationships built from long-term commitments, shared experiences); and, *monetary resources* (e.g. monetary support during family crises) (see Kahne & Bailey, 1997).

Turned on its head, however, there is much less to be found regarding the impact of place, inwardly, upon the schools. Dryfoos (1999) does document a growing trend to expand the usages of school facilities around community needs (e.g. before- and after-school programming, senior-centers, community "hubs," youth recreation and arts centers). But, by far the most intriguing line of inquiry, to date, into the reactions of schools to "place," is coming out of a growing historical interest in the study of school-community relations during pre-desegregation times, in African American neighborhoods.

Carter Savage (1999), for example, presents the story of African American schooling in Franklin, Tennessee (1890–1967) – as a story of a school's deep sense of "agency" vis-à-vis its community. High expectations and a strict moral code were passed from residents to teachers, and back again. Stretching resources to their fullest but looking to the community for the supplemental brought parents actively into constant spaghetti suppers, "pie struts," and cake walks. Community volunteer energies were heavily invested at the school in sports, talent nights, the school band, proms, and pageants. "The school", concluded Savage (1999: 27), "belonged to the Black community." "Ownership gave them a control of their destiny."

Back and forth. Community-to-school, and school-to-community. We still know very little about how, reciprocally, place matters. For long, buffering schools from their communities, and viewing communities as collections of "deficits" above strengths, have been at the center of much administrative attention. These myopic perspectives have changed; but the question is still: How can place replace buffering as a much stronger and deeper theoretical construct?

Institutional Investments

A notion of investment exists at the heart of much current thinking in the community-development arena. Linkages have been established, in many neighborhood settings, with a federal program labeled EZ/EC (Empowerment Zones/Enterprise Communities). This is a national effort to encourage grassroots' action in community revitalization, through the provision of an array of incentives to spur public and private investment. As mentioned earlier, the very language of development, at present, is heavily laden with investment-type terminology (e.g. entrepreneuralism, enterprising, self-reliance, restoration, indigenous leadership, mobilization of resources).

Investment is also much to be found as a focusing idea in a rethinking of social capital formation. In their review of this literature, Driscoll and Kerchner (1999: 395–397) identified communities as "zones of production" – wherein the production and transmission of knowledge, often flowing heavily out of the work of the school, can be regarded as a "basic industry" of the city. Much the same line of thinking was developed in a "theory of the state" analysis by Claus Offe and Volker Ronge (1997). In place of a system of transfer payments (as in welfare or public housing subsidies), argue Offe and Ronge (1997), the new role of the state must be to *create the conditions* under which individuals and neighborhoods can achieve independent success. This means a program of "public infrastructure investment" in education and training, in community development, and in "general adaptivity" (Offe & Ronge, 1997: 62).

Another important idea in present-day community development, however, is that communities themselves are full of strengths and assets, and contain many of the conditions of their own success. There are sizeable assets especially to be found in an array of community institutions. This list includes the schools, of course, but it also includes the parks, community associations, libraries, local banks, hospitals, religious groups, the police, for-profit businesses, youth organizations, day care centers, and groupings of local residents (e.g. seniors).

A central task, noted Kretzmann and McKnight (1993) is how to identify and "capture" local institutions for community building. Many community institutions (and even the schools), warned Morris Janowitz as early as 1952, follow a policy of "limited liability" vis-à-vis their communities, wherein the institutions demand more from the community than they invest. Thus, 'capturing' can include assisting an institution in realizing its own stake in, and taking steps toward, the development of a neighborhood. There are varieties of powerful institutional roles among effectively "captured" institutions – from the direct investments of neighborhood banking or entrepreneurship, to the indirect but no less powerful investments of mobilization and coalition – building or assistance in "straddling" institutional boundaries toward cooperative endeavor (see Halpern, 1995; Mouritsen & Skaerbaek, 1995; Couto, 1998).

As with the concept of place, very little is known to date about how to get local institutions to move from a sense of limited to full liability or to effectively "straddle" in community development. Interestingly, one community institution that has been increasing markedly in the scope of its involvement in both community economic development and education is the local church – particularly in the form of a faith-based communities movement (spreading rapidly across the nation) (see Shirley, 1997). Out of this, there may be a blurring of institutional boundaries underway in some communities that would have been fully unexpected and perhaps even taboo just a few short years ago.

Empowerment

Empowerment is an extremely difficult construct to get a solid "handle" on; and it is easily misinterpreted or misconstrued. Errors in thinking can abound, in assumptions that to empower is to transfer some controls over decisionmaking from those persons with power to those without. To empower, assumably, is to supply someone anew with substantial and even forceful influence – or, if not fully that, to at least move from power-over to power-with (e.g. administrators *with* teachers, teachers *with* parents). A key error in such thinking is that there is a certain largesse implied – persons without are now *em*-powered, by other folks. There is a failure to recognize that much power is usually already there, even among the "weaker" participants in decision making.

A deeper sense of just what empowerment might mean comes to us from "regime theory" (Elkin, 1987; Stone, 1989; Ramsay, 1996). With roots in many decades of community studies, regime theory is a "culturist" interpretation of power – asserting that power as a key, structural element exists in a closely reciprocal relationship with each community's overall way of life (e.g. its essential culture, social institutions, local history, values, expectations, local markets, etc.). Interestingly, some early theorizing around regime came out of studies of third-world economic development, wherein powers and beliefs rooted in peasant cultural traditions were frequently found to interfere with the assumed "rationality" of economic incentives (Scott, 1976).

Regime theorizing can go either way within the old debate surrounding pluralism vs. elitism in politics. Indeed, Slater and Boyd (1999: 325) observe that in Greek philosophy the term regime could cover widely varied forms of a "polity" – from one individual ruling in the interests of the whole to either few or many ruling in the interests of the whole. Who has more power or less power may not be as important to an understanding of regime politics than just what is the expression of power to be found in a community's way of life, and certainly in the course of its revitalization.

Furthermore, regime theorizing can go either way in the old discussion around the "two faces" of politics (the power to change vs. the power to halt change) (see Bachrach & Baratz, 1962). It was Ramsay's (1996) observation that economic development and growth may frequently disrupt established power relations and upset delicately balanced neighborhood arrangements, thereby threatening many of the "powers-that-be" (see also Smith, 1998). On the other hand, cultures can be enormously powerful when well harnessed to and integrated with a community's values, concerns, and interests – when both the regime and forces for rejuvenation or development are following a fairly common script.

Won't happen, warns Seymour Sarason (1995: 13), unless one has managed first to "change the culture of schools in specific ways." Ah, but it can happen, counter Gary Wehlage et al. (1989), if there are supportive school structures emphasizing community engagement, reciprocity, and partnering. Joseph Kahne (1996: 129–136) would add the caveat that a politics of change or "reform" must begin with some regime-minded notions of social adjustment and social sensitivity – in short, with evidence of social capital development at the school's end of things, not just the community's.

Partnering

The idea of the "partnership" is currently hot stuff in the educational reform literature. School partnerships with parents and families, with non-profit community organizations, with the for-profit sector, with colleges and universities, with local health-service institutions, and with an array of additional people-serving agencies of government – comprise together the most common targets of partnering opportunity (see particularly Epstein, 1994; Cibulka & Kritek, 1996; Dorsch, 1998; Springate & Stegelin, 1999).

Interestingly, however, the partnership as a key tool of reform has received far from adequate examination. Even less attention to date has been paid to the role of partnering in community-development activities. The partnership as a construct is heavily value-laden – with appealing images of connection, collaboration, collegiality, and sharing. It is decidedly a good thing to do.

Nevertheless, both the management and the organizational theory literatures provide some food for thought. For example, Kraatz (1998) found that "weak" ties between partnering organizations can impede the ability of both partners to adapt effectively to environmental change. Alternatively, "strong" partnership ties help. The discovery additionally, though, was that strong partnering may require a

> . . . social similarity within a network [that] also promotes frequent communication, as well as liking and intimacy between actors (Kraatz, 1998: 624).

As a second addendum, on much the same point, Ullman (1998) discovered that a relationship is weakened if the *capacity* of any one partner is limited by bureaucratic inefficiency, rules inflexibility, or similar organizational constraints. Similarly, Gulati (1995: 619) found that a strong partnership is facilitated if there is information readily available "about the specific capabilities and reliability of potential partners."

Additional issues, thoroughly underexplored to date, might involve such partner-related elements as:

(1) *Ownership* – with an understanding that owners ranging from donors or philanthropists, to stockholders, to church members, to taxpayers may all differ considerably in motivations and definitions of client "need" (Schlesinger, 1998).

(2) *Rhetoric* – with an appreciation of the role that expressive outputs (from protest, to cultural affirmation, to agenda-setting, to emotional appeals) may play in cross-agency cooperation (Hage, 1998).

(3) *Mission* – particularly as the fulfillment of mission becomes separately wrapped into the resources and lifeways of persons serving organizations in volunteer vs. professional capacities (Karl, 1998). Finally,

(4) *Strategy* – appreciating the differing legal and resource foundations, environmental "mapping," internal conflicts, legitimacy concerns, etc. which together can influence how each organization may approach a shared developmental agenda (Alexander, 1998).

Specific to the educational arena, Knapp and Brandon (1998) have identified a useful set of organizational difficulties in promoting successful partnering. These include such "structural constraints") as differing vocabularies, differing ideas as to what knowledge is most useful, differing reward systems, and differences in respect for a clear hierarchy of authority. "Institutional constraints" are also of importance – in such potential-for-partnering domains as reliance upon professional expertise, the use of teamwork, the apportionment of credit or blame, and what is or is not regarded as a "problem" in each organization (Knapp & Brandon, 1998).

Again, however, the appealing idea of the partnership in cementing school to community is now only minimally understood. About all one can say at this point, concludes Nina Dorsch (1998: 196) from a case study, is that it takes time: "Time is necessary for trust to develop, for courage to grow, for communication to become open, for commitment to develop, and for vision to emerge."

Beyond time, add Sarason and Lorentz (1998) in a perceptive essay, the partnership concept will decidedly need a push. One approach to the push, they suggest, could well be a new role for schools of partnership "coordinator." This would be a full-time job scanning the neighborhood for resource-exchange opportunities, forging a network of those persons and agencies whose self-interests would be furthered by partnering, and doing whatever else it takes to 'cross boundaries.'

THE CHAPTERS TO FOLLOW

In brief summary, a new thrust is underway toward linking and of course under-standing the combined developmental fortunes of both neighborhood and school. In community-development theorizing, the improvement of the school and the revitalization of the community are part of a common agenda – an agenda with some possibly instructive theoretical insights into the importance of place, just what it means to invest and to empower, and what might be gained in school improvement through effective partnering. The chapters collected in this volume represent, together, a very early foray into the theorizing and problem-finding that will be required – if a shared revitalization agenda is to be realized.

The collection begins with some further theorizing, in depth, by Mary Erina Driscoll on the concept of "sense of place." Driscoll explores the role of place in education's history, probes the extant literature for a definition of sense-of-place, pulls some central dimensions of a sense of place from that literature, and proceeds toward a set of key implications for neighborhood schools. Driscoll suggests the utility for public education of a "re-imagined" neighborhood school – a school that fully appreciates place and its importance, commits itself to extending the place-related boundaries of the school, has an awareness of the shared social construction of the neighborhood and the school, and understands "that it is only when we know and live in an educational place that we may be prepared to leave it."

The chapter contributions continue with further case-based theorizing by Charles Kerchner and Grant McMurran. The authors explore the unique example of a community-development – intended investment – by, quite surprisingly, the public school district serving Pomona, California (Pomona Unified). A dying commercial mall was redirected by the district into a variety of educational and family-supportive services. An effect, as a spin-off, was the return of other investments and resources to central Pomona, as well as a return of children-laden families and repaired housing. In theorizing around the Pomona case, Kerchner and McMurran expand upon some earlier analyses of the productive capacities of schools-productive, that is, in helping to foster and guide community-development activities (Driscoll & Kerchner, 1999). The key theoretical notion in this chapter is of "schools as engines" – articulating a new appreciation of opportunities for shared growth and development between school district and community.

However, the chapter contributed by Robert Pena, Cristal McGill, and Robert Stout warns us that a newly-designed "engine" of development-oriented schooling can easily derail. Their case study from Arizona discusses strains and contradictions between academic and non-academic (or family-support) agendas, the new range of problems for schools that can emerge from the severity and unpredictability of needy communities, and the limited capacity of communities to revitalize without a sizeable infusion of external resources. Perhaps of most concern, note the authors, is the observation that to work effectively with communities, a service-oriented profession must "open one's soul to examination." But, to do so and see little immediate effect can enhance doubt and be enormously discouraging.

The challenges of a community-development role for the schools are even more broadly articulated in the contribution by Rodolfo Torres and Louis Miron. You cannot work effectively in most communities, these chapter-authors suggest, without a much deeper sense of culture and its consequences that has typically shaped the work of professional educators. Indeed, in their examination of the Latino experience in Los Angeles, Torres and Miron claim that not only a lack of understanding but preconceived notions of culture can be constraining. Furthermore, understandings are destructive that are not fully, larger-context cognizant of Latino workforce experiences, the economic and political strains of social class inequalities, and the isolations of class structures. Not just an improved sense of place but a deeper appreciation for 'the dialectics of landscape' must begin to drive development, conclude Torres and Miron.

The contribution by Charis McGaughy begins a grouping of chapters with reports of on-site studies of community-development experimentation. McGaughy examines a federally-funded effort which begin in 1994, in Akron, Ohio. This is an EZ/EC (Empowerment Zone/Enterprise Community) project considered to be somewhat unique – in that it has included, from-the-beginning, a partnership with public education. The Akron program has demonstrated a solid sense of conceptual underpinnings, as well as an "amazing alignment" of participants' goals, notes McGaughy. However, she goes on, there appears to be less "buy-in" on the business side to social and community-ecology elements in the project against the economic and job-training side of the equation. The "community-oriented good works thing" that appeals most to educators, continues McGaughy, just doesn't come close to "something like Tech Prep" in interesting the business community.

Impressive success to not-quite-there-yet is also the story summarized in the chapter by Dennis Shirley. Shirley analyzes the community-organizing dimension of development/reform in two schools affiliated with the work of the Industrial Areas Foundation (IAF), in Texas. Of major importance in Shirley's work is a focus of attention inward, upon *school-reform*. This is the turned-on-its-head component of social capital development – where schools can learn to "make the most of the social capital in the community" and "community organizing becomes a central part of the culture of a public school." While a more successful story at just one of his two case-study schools, Shirley's analysis, again, is unique in providing a direct look at community-effects *upon the schools*, rather than the other way around.

Community-effects upon the schools could be enormously facilitated, argue Goldring and Hausman in the next chapter – *if* local principals can begin building "civic capacity" in their respective neighborhoods. Civic capacity goes well beyond the notion of social capital development; it is a blending of outreach, investment, and partnering – wherein principals actively join forces with parents and other agencies in focused community-engagement efforts. Unfortunately, continue these chapter-authors, a survey of principals in two large urban districts shows little current involvement with community agencies and a continued focus of attention inward (e.g. upon classroom instruction), even when principals are offered an hypothesized choice of changed priorities.

The role of the principal as builder of civic capacity will not come easily, warn Goldring and Hausman, amid a heavy array of competing time-demands, accountability pressures, and shortfalls in administrator preparation.

Interestingly, notions of civic capacity as well as community-effects achieve a unique twist in the following chapter, contributed by Claire Smrekar. Smrekar examines the growing phenomenon of "workplace schools" – where corporate interests in serving the family educational needs of their employees drive quite a different approach to key questions of place and partnership. The community of geographic residence, argues Smrekar, is not necessarily the community of greatest time-allocation or even of deepest psychological meaning for many of today's hardworking families. Schools at the workplace can have a "choice" flavor and, moreover, can offer a new route toward the social integration of work, family, and school.

Furthermore, not all but some workplace schools can even serve community-development goals, as discovered by Smrekar in her case-study of the Des Moines (Iowa) Downtown School. This workplace school serves a number of city-center employers but it also represents a business/education alliance that is designed to help revitalize and "bring the community back" to downtown Des Moines. Questions abound, of course, including: Just what is the nature and the pedagogical impact of a reconstituted community around workplace, in lieu of residence, in American education?

Finally, the book closes with some wrap-it-up theorizing by Hanne Mawhinney, around central issues in merging the processes of development and civic engagement. Within the context of some new directions in policymaking for the school-to-work transition, there is now an opportunity, argues Mawhinney, to explore deeply the contributions of schools to a meaningful social/economic integration of their surrounding communities. Schools "do not bowl alone;" their activities and partnerships can help re-invigorate "processes of development build[ing] networks of relationships that in turn foster the kind of civic engagement that focuses on the common good of the entire community."

A BRIEF SUMMARY

From a rediscovery of "place" and even workplace in schooling, to the school as an "engine" of development in its community, to issues in *community*-driven school reform, to a broadening of the social capital construct into notions of "civic capacity" and "civic engagement" – there is much in this collected set of papers that introduces exciting new directions in thinking through school-improvement issues.

Place and localities matter-reflecting now a sense of the deep embeddedness of schools in their communities in ways that go far beyond earlier, schools-reaching-out notions of the relationship. The community-reaching-in side is now an additionally recognized key to potentialities for school reform. Sharing and reciprocity rather than bridging and buffering; partners in community growth and development; "opening one's soul" to community examination and embracing the community's culture; joining other community institutions in both empowerment and investment; merging school, community, and workplace in an expanded domain of developmental activity; partnering with community institutions in school-to-work; and, finally, in school administration, moving effectively toward the doing-it-together construction of "civic capacity" and "civic engagment."

Each of the above, key ideas in this set of chapters is just-a-beginning. Much inquiry lies ahead. In the arena of urban development generally, neighborhoods are just now being rediscovered, indeed, revitalization block-by-block is still more vision than reality – but it is a paradigm rapidly gaining national attention and much investment (Sherman, 2000). The paradigm is also just beginning to display a realization that community-development has a vital and necessary link with schooling; moreover, *every* community's wide array of resources can be enormously educative.

REFERENCES

Alexander, V. D. (1998). Environmental Constraints and Organizational Strategies: Complexity, Conflict, and Coping in the Nonprofit Sector,' In: W. W. Powell & E. S. Clemens (Eds), *Private Action and the Public Good* (pp. 272–290). New Haven: Yale University Press.

Bachrach, P., & Baratz, M. S. (1962). Two Faces of Power. *American Political Science Review*, *56*, 947–952.

Behrman, R. E. (1999). When School Is Out. *The Future of Children*, 9(2).

Booth, A., & Dunn, J. F. (1996). *Family-School Links: How Do They Affect Educational Outcomes?* Mahwah, NJ: Lawrence Erlbaum Associates.

Boyd, W. L. (1996). *Competing Models of Schools and Communities: The Struggle to Reframe and Reinvent their Relationships.* Invited Keynote Address for a Conference on Leading the Learning Community, Australian Council for Educational Administration, Perth, Western Australia.

Boyd, W. L., Crowson, R. L., & Gresson, A. (1997). Neighborhood Initiatives, Community Agencies, and the Public Schools: A Changing Scene for the Development and Learning of Children. In: M. C. Wang & M. C. Reynolds (Eds), *Development and Learning of Children and Youth in Urban America* (pp. 81–99). Philadelphia: Temple University Center for Research in Human Development and Education.

Cibulka, J. G., & Kritek, W. J. (1996). *Coordination Among Schools, Families, and Communities*: Prospects for Educational Reform. Albany: State University of New York Press.

Coleman, J. (1988). Social Capital and the Creation of Human Capital. *American Journal of Sociology, 94*, Supplement, S95–S120.

Couto, R. (1998). Community Coalitions and Grassroots Policies of Empowerment. *Administration & Society, 30*(5), 569–594.

Crowson, R. L. (1999). *The Turbulent Policy Environment in Education.* Paper presented at the Annual Meeting of the American Educational Research Association (AERA), Montreal.

Crowson, R. L., & Boyd, W. L. (1993). Coordinated Services for Children: Designing Arks for Storms and Seas Unknown. *American Journal of Education, 101*(2), 140–179.

Dorsch, N. G. (1988). *Community Collaboration and Collegiality in School Reform: An Odyssey Toward Connections.* Albany: State University of New York Press.

Driscoll, M. E., & Kerchner, C. T. (1999). The Implications of Social Capital for Schools, Communities, and Cities, In: J. Murphy & K. S. Louis (Eds), *Handbook of Research on Educational Administration* (2nd ed.) (pp. 385–404). San Francisco: Jossey-Bass.

Dryfoos, J. G. (1999). The Role of the School in Children's Out-of-School Time. *The Future of Children, 9*(2), 117–134.

Elkin, S. L. (1987). *City and Regime in the American Republic Chicago:* University of Chicago Press.

Fuller, W. E. (1987). *The Old Country School: The Story of Rural Education in the Middle West.* Chicago: University of Chicago Press.

Garr, R. (1995). *Reinvesting in America.* Reading, MA: Addison-Wesley.

Gulati, R. (1995, December). Social Structure and Alliance Formation Patterns: A Longitudinal Analysis, *Administrative Science Quarterly, 40*(4), 619–652.

Hage, J. (1998). Reflections on Emotional Rhetoric and Boards for Governance of NPO's. In: W. W. Powell & E. S. Clemens (Eds), *Private Action and the Public Good* (pp. 291–301). New Haven: Yale University Press.

Halpern, R. (1995). *Rebuilding the Inner City: A History of Neighborhood Initiatives to Address Poverty in the United States.* New York: Columbia University Press.

Haveman, R., & Wolfe, B. (1994). *Succeeding Generations: On the Effects of Investments in Children.* New York: Russell Sage Foundation.

Havighurst, R. J. et al. (1962). *Growing Up in River City.* New York: John Wiley & Sons, Inc.

Janowitz, M. (1952). *The Community Press in an Urban Setting: The Social Elements of Urbanism.* Chicago: The University of Chicago Press.

Kahne, J. (1996). *Reframing Educational Policy: Democracy, Community, and the Individual.* New York: Teachers College Press.

Kahne, J., & Bailey, K. (1997). *The Role of Social Capital in Youth Development: The Case of I Have a Dream.* A Paper presented at the Annual Meeting of the University Council for Educational Administration (UCEA), Orlando.

Karl, B. D. (1998). Volunteers and Professionals: Many Histories, Many Meanings. In: W. W. Powell & E. S. Clemens (Eds), *Private Action and the Public Good* (pp. 245–257). New Haven: Yale University Press.

Kerchner, C. T. (1997). Education as a City's Basic Industry. *Education and Urban Society, 29*(4), 424–441.

Kerchner, C. T., Koppich, J. E., & Weeres, J. G. (1997). *United Mind Workers: Unions and Teaching in the Knowledge Society.* San Francisco: Jossey-Bass.

Knapp, M. S., & Brandon, R. N. (1998). Building Collaborative Programs in Universities, In: M. S. Knapp and Associates (Eds), *Paths to Partnership* (pp. 139–164). Lanham, MD: Rowman & Littlefield Publishers, Inc.

Kraatz, M. S. (1998). Learning By Association? Interorganizational Networks and Adaptation to Environmental Change. *Academy of Management Journal, 41*(6), 621–643.

Kretzman, J. P., & McKnight, J. L. (1993). *Building Communities From the Inside Out: A Path Toward Finding and Mobilizing a Community's Assets*. Evanston, IL: Institute for Policy Research, Northwestern University.

Leithwood, K., & Jantzi, D. (1999). The Relative Effects of Principal and Teacher Sources of Leadership on Student Engagement with School. *Educational Administration Quarterly, 35,* Supplemental, 679–706.

Littell, J., & Wynn, J. (1989). *The Availability and Use of Community Resources for Young Adolescents in an Inner-City and a Suburban Community*. A Report. Chicago: University of Chicago, Chapin Hall Center for Children.

Mouritsen, J., & Skaerback, P. (1995). Civilization, Art, and Accounting: The Royal Danish Theater- An Enterprise Straddling Two Institutions, In: W. W. Scott & S. Christensen (Eds), *The Institutional Construction of Organizations* (pp. 91–112). Thousand Oaks, CA: Sage Publications.

Offe, C., & Ronge, V. (1997). Thesis on the Theory of the State, In: R. E. Goodwin & P. Petitt (Eds), *Contemporary Political Philosophy: An Anthology* (pp. 60–65). Cambridge, MA: Blackwell Publishers, Ltd.

Ogbu, J. U. (1974). *The Next Generation: An Ethnography of Education in an Urban Neighborhood*. New York: Academic Press.

Peshkin, A. (1997). *Places of Memory: Whiteman's Schools and Native American Communities*. Mahwah, NJ: Lawrence Erlbaum Associates.

Ramsay, M. (1996). *Community, Culture, and Economic Development: The Social Roots of Local Action*. Albany: SUNY Press.

Sarason, S. B. (1995). *Parental Involvement and the Political Principle: Why the Existing Governance Structure of Schools Should Be Abolished*. San Francisco: Jossey- Bass.

Sarason, S. B., & Lorentz, E. M. (1998). *Crossing Boundaries: Collaboration, Coordination, and the Redefinition of Resources*. San Francisco: Jossey-Bass.

Savage, C. J. (1999). *Because We Did More With Less: The Agency of African American Teachers in Franklin, Tennessee, 1890–1967*, A Paper presented at the Annual Meeting of the American Educational Research Association (AERA), Montreal.

Schlesinger, M. (1998). Measuring the Consequences of Ownership: External Influences and the Comparative Performance of Public, For-Profit, and Private Nonprofit Organizations, In: W. W. Powell & E. S. Clemens (Eds), *Private Action and the Public Good* (pp. 85–113). New Haven: Yale University Press.

Schorr, L. B. (1997). *Common Purposes: Strengthening Families and Neighborhoods to Rebuild America*. New York: Anchor Books.

Scott, J. C. (1976). *The Moral Economy of the Peasant*. New Haven: Yale University Press.

Sherman, L. (2000). What's Wrong with America Revitalizing Urban Communities Block by Block, *Journal of City and State Public Affairs*, Vol. 1, No. 1, 1–6.

Shirley, D. (1997). *Community Organizing for Urban School Reform*. Austin: University of Texas Press.

Shuman, M. H. (1998). *Going Local: Creating Self-Reliant Communities in a Global Age*. New York: The Free Press.

Slater, R. O., & Boyd, W. L. (1999). Schools as Polities. In: J. Murphy & K. S. Louis (Eds), *Handbook of Research on Educational Administration* (2nd ed.) (pp. 323–335). San Francisco: Jossey-Bass.

Smith, B. H. (1998). Non-profit Organizations in International Development: Agents of Empowerment or Preservers of Stability? In: W. W. Powell & E. S. Clemens (Eds), *Private Action and the Public Good*. New Haven: Yale University Press.

Smrekar, C. E., & Mawhinney, H. B. (1999). Integrated Services: Challenges in Linking Schools, Families, and Communities, In: J. Murphy & K. S. Louis (Eds), *Handbook of Research on Educational Administration* (2nd ed.) (pp. 443–461). San Francisco: Jossey-Bass.

Stone,. C. N. (1989). *Regime Politics: Governing Atlanta,1945–1988.* Lawrence: University of Kansas Press.

Ullman, C. F. (1998). Partners in Reform: Nonprofit Organizations and the Welfare State in France, In: W. W. Powell & E. S. Clemens (Eds), *Private Action and the Public Good* (pp. 163–176). New Haven: Yale University Press.

Vandell, D. L., & Shumow, L. (1999). After-School Child Care Programs. *The Future of Children, 9*(2), 64–80.

Wehlage, G. G., Rutter, R. A., Smith, G. A., Lesko, N., & Fernandez, R. R. (1989). *Reducing the Risk: Schools as Communities of Support.* London: The Falmer Press.

White, J. A., & Wehlage, G. (1995). Community Collaboration: If It Is Such a Good Idea, Why Is It So Hard To Do? *Educational Evaluation and Policy Analysis, 17*(1), 23–38.

THE SENSE OF PLACE AND THE NEIGHBORHOOD SCHOOL: IMPLICATIONS FOR BUILDING SOCIAL CAPITAL AND FOR COMMUNITY DEVELOPMENT

Mary Erina Driscoll

INTRODUCTION

At the beginning of his autobiography *The Education of Henry Adams*, the author renders two poignant, powerful descriptions of the places that shaped his childhood and his education. In each, the physical characteristics of the setting are intertwined with recollections of the activities that occupied him in that locale. In these accounts Adams relates sense memories that bring to the reader vivid details of the city and country of his youth; these contrasting pictures become metaphors for the ways in which he has come to know and experience life. The Boston of his boyhood, he tells us, "was winter confinement, school, rule, discipline; straight gloomy streets, piled with six feet of snow in the middle; frosts that made the snow sing under wheels or runner; thaws when the streets became dangerous to cross . . ." Boston is also remembered as a "*society* of uncles, aunts, and cousins who expected children to behave themselves, and, who were not always gratified . . ." (Italics mine) The spiritual and philosophical constraints created by this sense of place are

Community Development and School Reform, Volume 5, pages 19–41
Copyright © 2001 by Elsevier Science Ltd.
All rights of reproduction in any form reserved.
ISBN: 0-7623-0779-X

real: ". . . Above all else, winter represented the desire to escape and go free. Town was restraint, law, unity" (Adams, 1918/1961, p. 8).

Adams contrasts this winter place with the rural Quincy where he spent his summers: "Country, only seven miles away, was liberty, diversity, outlawry, the endless delight of mere sense impressions given by nature for nothing, and breathed by boys without knowing it. . . ." (Adams, 1918/1961, p. 8). So many sense impressions combine, he tells us, that "To the boy Henry Adams, summer was drunken. Among senses, smell was the strongest – smell of hot pine-woods and sweet-fern in the scorching summer noon; of new-mown hay; of ploughed earth; of box hedges; of peaches, lilacs, syringas; of stables, barns, cow-yards; of salt water and low tide on the marshes . . ." The sense of place also comes from visual experience, a perspective he fully appreciates only years later: "Light, line and color as sensual pleasures, came later and were as crude as the rest. The New England light is glare, and the atmosphere harshens color. The boy was a full man before he ever knew what was meant by atmosphere; his idea of pleasure in light was the blaze of a New England sun. His idea of color was a peony, with the dew of early morning on its petals. The intense blue of the sea, as he saw it a mile or two away, from the Quincy hills; the cumuli in a June afternoon sky . . ." (Adams, 1918/1961, p. 8).

Boston and Quincy symbolize two different aspects of his nature, Adams reports: "The bearing of the two seasons on the education of Henry Adams was no fancy; it was the most decisive force he ever knew . . ." (Adams, 1918/1961, p. 9). This duality becomes critical in helping him understand the world: "Winter and summer, cold and heat, town and country, force and freedom, marked two modes of life and thought, like balanced lobes of the brain." (Adams, 1918/1961, p. 7) In his memory and his education, the effects of these places persist. As he concludes:

"Winter and summer, then, were two hostile lives, and bred two separate natures. Winter was always the effort to live; summer was tropical license. Whether the children rolled in the grass, or waded in the brook, or swam in the salt ocean, or sailed in the bay, or fished for smelts in the creeks, or netted minnows in the salt-marshes, or took to the pine-woods and the granite quarries, or chased muskrats and hunted snapping-turtles in the swamps, or mushrooms or nuts on the autumn hills, summer and country was always compulsory learning. Summer was the multiplicity of nature; winter was school" (Adams, 1918/1961, p. 9).

Henry Adams' keen articulation of a sense of place reminds us that the settings for our education and growth are potent influences on the people we become. The development of a sense of place in and around the educational settings we construct is thus an important tool in understanding and improving our schools. In this chapter, I explore the ways in which a sense of place can inform our imagination, especially as we re-envision our traditional "neighborhood school" amidst a policy climate that seems to move inexorably towards national standardization. First, I will argue that an appreciation of a local sense of place in our schools has fallen in some disfavor, and that reclaiming this sense of place for our educational discourse has some merit. Second, I will articulate what I mean by a sense of place, borrowing from a broadly-based literature. Third, I will link this construct of place to some earlier work that explored the ways in which schools can help to develop social capital (Driscoll & Kerchner, 1999). Finally, I will use the sense of place to explore some ideas about the neighborhood school and its role in community development.

RECLAIMING A SENSE OF PLACE

Schools have always played two roles in American society. On the one hand, they are linked to the familiar places of home, family and community; on the other, they stand as the primary institution charged with the preparation of children for the 'broader' worlds of society, polity and nation.

This central dilemma of American education is beautifully rendered by William Proefriedt in an article entitled "Education and Moral Purpose: The Dream Recovered." Proefriedt (1985) remarks on the critical role that schools played in the twentieth century as a means toward achieving the American dream of success. That dream, with its promise of equality of opportunity for all, ". . . has usually been located in the future and in the wide world out there, and yet at the same time we have seen the importance of the small town, of the local community, of one's childhood, one's past, to the fulfillment of the dream.' (Proefriedt, 1985, p. 406.) Like the dream itself, "The schools, too, express, this ambiguity. They are asked to play two often conflicting functions: to transmit the values of the immediate society, the small town, the local or ethnic group, to the children, and, at the same time, to introduce them to a set of skills and values that will enable them to gain entrance to the wider world, to make their way into the future, to fulfill the dream." (Proefriedt, 1985, pp. 406–407.) In sum, "The school is expected, and sometimes even by the same people, to teach the boy to live in the town, to stay off the railroad tracks, and at the same time, to put him on the train, to point him in the direction of the city." (Proefriedt, 1985, p. 407).

For at least the last half of the twentieth century the importance of the local perspective and the celebration of the "small town" was de-emphasized in favor of the development of a national educational policy that is designed to diminish the differences across state and local settings. James Bryant Conant's The *American High School Today* (1959), for example, argued for a national blueprint for comprehensive high schools that had powerful effects on the ways schools were organized for the ensuing three decades. For Conant, the small, local school was indeed the enemy – a constricting setting that failed to provide equal opportunities for all students. His assessment of the needs of the time, however, were colored by the tumultuous post-War climate that he believed justified new educational priorities. His report was written only five years after the landmark civil rights decision by a Supreme Court determined to overturn local preferences for segregated schools. It was published at the height of post-Sputnik fervor that energized the educational community with national visions of bolstered defense in face of international peril. Conant helped to shift educational priorities away from a consideration of the merits of the local and towards a renewed emphasis on the importance of schools in a national system of education oriented toward larger societal priorities. The structural and organizational forms that this emphasis took on in the ensuing years of educational expansion during the baby boom ensured that large and often regional institutions dominated the American educational market for the last part of the century.

There are many good reasons that we have moved away from locally flavored schools to an education system that is influenced by policy at the national level. Whether in the 1957 post-Sputnik calls for improved defense through better education or in our present day, the movement to make education a national priority, as opposed to a local one, has argued for more uniform standards of educational opportunity and achievement. This effort is laudable. No fondness for the local community can allow us as a society to be tolerant of pernicious differences across state and local systems that systematically deprive students of opportunities as a result of a geographic accident of birth. Similarly, the romantic view of homogenous and welcoming small towns and communities neglects the harsh realities of xenophobia, bigotry, racism and religious prejudice that too often lurked just beneath the surface.

But a reconsideration of a sense of place in education does not mean that we no longer aspire to the best education for all students, regardless of where they are located. Rather, by reconsidering a sense of place, we are admitting that there is no monolithic solution to the educational enterprise. Especially as we become a more diverse, multicultural, multiethnic society, we need to explore the many ways in which educational excellence may be achieved. To return to the Proefriedt/Wolfe metaphor for a moment, we are aware that we neglect an important piece of the educational equation if we think about schools only as the train that moves the child away from home and community to the city beyond. Consideration of those contemporary schools that seem to "work" points our attention in another direction – to the examination of the locally created, sustainable institutions that are intimately connected with the communities in which they find themselves as a means of preparing students to take on the broader world. We want to imagine the possibilities of schools that do not see themselves succeeding in spite of the community, but, rather, that envision themselves as key institutional players in the development of community. Such schools help to create the kind of social capital that supports and sustains growth and education.

A century ago, educational administrators and professors were occupied with the development of a science of education that melded the scientific management of Taylor with the universalistic understandings of educational psychology and assessment of Judd and Thorndike. But, as many have noted, that effort was doomed to failure, and there is not, indeed, "one best system" (Tyack, 1974; Tyack & Hansot, 1982). We seek instead to imagine what schools, teachers and students would do if educational administration were constructed "as if a sense of place mattered." (Driscoll & Kerchner, 1999).

It is no accident that Henry Adams' recollection of school, in contrast with his less formal education, was dismal: "most school experience was bad," he tells us; "the happiest hours of the boy's education were passed in summer lying on a musty heap of Congressional Documents in the old farmhouse at Quincy, reading . . . and raiding the garden at intervals for peaches and pears. On the whole he learned most then" (Adams, 1918/1961, p. 39).

By reclaiming a sense of place, we hope to move away from standardized, civilized "Boston" notions of education to embrace Quincy, and the "multiplicity of nature" that Adams celebrates and finds so useful.

THE MERITS OF A SENSE OF PLACE FOR UNDERSTANDING EDUCATION

Some of those who study place argue that only by understanding and sustaining our connections with the particular worlds of our experience can we grow to a more global understanding of the world at large and its needs. As William Leach argues, "A strong sense of place, along with the boundaries that shape it and give it meaning, not only fosters creativity but also helps to provide people – especially children – with an assurance that they will be protected and not abandoned . . . It is indisputable that children need a sense of place (along with an acceptance of boundaries that define and establish the safeness of place) in order to become self-reliant" (Leach, 1999, p. 179).

Moreover, continues Leach, these connections are essential if we are to build a productive and compassionate society:

> Without a sense of boundaried place, finally, there can be no citizenship, no basis for common bonds to others, no willingness to give to the commonwealth or to be taxed, even lightly, on behalf of the welfare of others. A living sense of a boundaried place, some kind of patriotism beyond love of abstract principles, is the main condition for citizenship. . . This living sense always has a provincial character. It takes shape first as connections to families and friends, then to neighborhoods, towns, and regions, and finally, to the nation and the world. It is through the formation of this sense of place, beginning with the home and parents, that people develop their loyalty to place, but it is only after the earliest concrete ties are formed that the bigger connections can be forged; the process cannot begin the other way around (Leach, 1999, pp. 179–180).

I would like to suggest a few other reasons that it may be profitable to consider a sense of place in conjunction with our current attempts to improve education. The first concerns the potential of these understandings to generate insights, drawn from across several disciplines, that may provide a useful counterpoint to our dominant assumptions about schools and school systems. The language of efficiency and standardization has long been our native tongue as we consider these institutions. The image of schools and their communities as rich and unique places, with deep local connections and singular sets of resources, cannot help to enrich our discourse. In this respect the development of a sense of education is similar to the re-envisioning of schools as communities that has become so popular over the past several years. Such a metaphor can provide a powerful alternative image to the bureaucratic icons of schooling with which we have become so familiar.

Second, there is something about the physicality or the geographic nature of place understandings that can be especially helpful as we consider schools today. Considering these geographic dimensions in their own right may have something to offer as we face the challenge of rebuilding a system that is in many cases characterized by crumbling, outdated and inappropriate physical facilities, too often located in deteriorating communities. Our attention to this dimension of school and of education is vital at this junction, and to reconnect with a sense of place may hold the promise of some helpful insights relative to this problem.

Third, I think that a serious consideration of place in American education holds the promise for an illuminating cultural critique of the places where we school our young people. Feminist geographers such as Janice Monk, for example (Monk, 1992), have demonstrated that the social experiences and histories of women are not recognized or celebrated in the same way as are those of men in the "gendered" public spaces (Spain, 1992) of many communities. Such an oversight sends an important cultural message. Similarly, careful attention to the suburban places where many women work as homemakers shows that the public construction of space and transportation often privileges those who work full time outside the home over the habits and needs of those attending to children full time (Winchester, 1992). These studies of place help us decipher some of the values we express in our public spaces and awaken us to the kind of respect these spaces convey for all who live and work there. Similarly, a careful study of the places we educate children should tell us if the messages we convey through them are consistent with the educational values that we preach.

HOW IS A SENSE OF PLACE DEFINED?

Examination of Place as a concept and efforts to articulate what is meant by a sense of place have occupied scholars across many different disciplines in recent years. Casey (1993), while citing the contributions of philosophers such as Heidegger and Bachelard to the philosophical study of place, notes as well the interest of "ecologically-minded geographers who attempted to reinstate place as a central category within their own discipline. A recent undercurrent of architects, sociologists, anthropologists, ethicists and theologians, feminists, and social observers is still gathering force . . ." (Casey, 1993, p. xv).

Tony Hiss (1990), in his *Experience of Place*, states simply: "The places where we spend our time affect the people we are and can become . . . Whatever we experience in a place is both a serious environmental issue and a deeply personal one" (Hiss, 1990. p. xi). He notes that "A brand-new science of place, growing up out of a body of formal research, like William H. Whyte's studies of plazas, is examining housing projects, train stations, hospitals, and sealed and sometimes "sick" office buildings; parks, lawns and traffic-clogged streets; entrances, steps, and views from windows; meadows, fields, and forests; light, colors, noises, and scents; the horizon; small-air ions, and wind speed; and privacy. "Among those who are engaged in this investigation," he tells us are "public health physicians, management consultants, architects, planners, clinical psychologists, ecologists, environmental psychologists, nature writers, political scientists, preservationists, and filmmakers. Although most of these students of place are still working separately, they have a common interest – safeguarding, repairing, and enriching our experience of place" (Hiss, 1990, pp. xv–xvi).

Even a cursory catalogue of these perspectives gleans a rich array of meanings for "a sense of place". For the sociologist David Hummon, sense of place is interwoven with notions of community satisfaction, community attachment, and community identity:

By *sense of place*, I mean people's subjective perspectives of their environments and their more-or-less conscious feelings about those environments (Steele, 1981). Sense of place is inevitably dual in nature, involving both an interpretive perspective *on* the environment and an emotional reaction *to* the environment . . . Whatever the balance of emotional and cognitive components, sense of place involves a personal *orientation* toward place, in which one's understandings of place and one's feelings about place become fused in the context of environmental meaning (Hummon, 1992, p. 262).

The cultural geographers Kay Anderson and Fay Gale emphasize both the territorial nature of place and the degree to which place is socially constructed. They argue that "(t)he cultural process by which people construct their understandings of the world is an inherently geographic concern. In the course of generating new meanings and decoding existing ones, people construct spaces, places, landscapes, regions and environments" (Anderson & Gale, 1992, p. 4).

The anthropologist Katherine Platt emphasizes the generative elements of place, along with its capacity for historical imagination:

Places capture experience and store it symbolically. Its collective meanings are extractable and readable by its later inhabitants. This symbolic housing of meaning and memory gives place temporal depth. But not only do places of experience store meaning about the past; they also are platforms for visions and plans about the future. Places of experience provision us with identity to venture forth out of this place into less certain or orderly spaces. Places of experience provide categories for managing new adventures and new cycles of old adventures. Places of experience connect the past to the future, memory to expectation, in an invigorating way. Places of experience give us a sense of continuity and energy (Platt, 1996, p. 112).

The naturalist John Brinckerhoff Jackson discusses the ways in which self and space interact to create a sense of place:

Most of us, I suspect, without giving much thought to the matter, would say that a sense of place, a sense of being at home in a town or city, grows as we become accustomed to it and learn to know its peculiarities. It is my own belief that a sense of place is something that we ourselves create in the course of time. It is the result of habit or custom. But others disagree. They believe that a sense of place comes from our response to features which are *already* there – either a beautiful natural setting or well-designed architecture. They believe that a sense of place comes from being in an unusual composition of spaces and forms – natural or man-made (Jackson, 1994, p. 151).

Jackson notes wryly that sense of place "is a much used expression, chiefly by architects but taken over by urban planners and interior decorators and the promoters of condominiums, so that it now means very little" (Jackson, 1994, p. 157). But he also discusses at length the origin of the term, which he terms "an awkward and ambiguous modern translation of the Latin term *genius loci*." He continues:

In classical times it meant not so much the place itself as the guardian divinity of that place. It was believed that a locality – a space or a structure or a whole community – derived much of its unique quality from the presence or guardianship of a supernatural spirit. The visitor and the inhabitants were always aware of that benign presence and paid reverence to it on many occasions. The phrase thus implied celebration or ritual, and the location itself acquired a special status (Jackson, 1994, p. 157).

"We now use the current version to describe the *atmosphere* to a place, the quality of its *environment*," he concludes (Jackson, 1994, p. 157).

Jackson also remarks that few people in modern times have any sense of place that is related to political space, although, at least until recently, he argues,

... what we have had are spaces and events related to *family* and the small neighborhood group. By that I mean not merely the home – which in the past was the basic example of the sense of place – but also those places and structures connected with ritual and with restricted fellowship or membership – places which we could say were extensions of the dwelling or the neighborhood: the school, the church, the lodge, the cemetery, the playing field (Jackson, 1994, p. 158).

For others, articulation of a sense of place is impelled by the perception of loss. In his recent book *Country of Exiles* (1999), William Leach starts from the assumption that there has been a "weakening of place as a centering presence in the lives of ordinary people." His book is "animated by the premise that the well-being of most Americans rests on a healthy connectedness to place, and that a wearing away of such a relationship is dangerous" (Leach, 1999, pp. 6–7).

For Leach, defining a sense of place also means discussion of what it is not. "I do not equate place with community," he writes, "because in recent years community has come to mean practically any group of people joined together by almost any shared characteristic (corporate, academic, racial, ethnic, sexual, and so forth). Community has been transformed into a transparent condition, barely related to concrete geographical places with histories . . ." (Leach, 1999, p. 7). He continues: "Place, of course, may contain or signify all these things – community, nature, property – in some measure, but its meaning is bound to a geographic reality both historical and profoundly lasting . . . At its best, it is the collective outgrowth of our control over our own lives and destinies" (Leach, 1999, p. 7).

The philosopher Edward Casey (1993) has undertaken extended studies of the role of place in philosophical systems. Its importance, he argues, is undeniable:

... It remains the case that where we are – the place we occupy, however briefly – has everything to do with what and who we are (and finally, that we are). This is so at the present moment: where you are right now is not a matter of indifference but affects the kind of person you are, what you have been doing in the past, even what you will be doing in the future. Your locus deeply influences what you perceive and what you expect to be the case (Casey, 1993 pp. xiii).

Post modern society has lost this connection to place, Casey argues, suffering much as a consequence:

> Rushing from place to place, we rarely linger long enough in one particular place to savor its unique qualities and its local history. We pay a heavy price for capitalizing on our basic animal mobility. The price is the loss of places that can serve as lasting scenes of experience and reflection and memory.
> ... As J.J. Gibson reminds us ... "We do not live in 'space.' Instead, *we live in places.* So it behooves us to understand what such place-bound and place-specific living consists in. However lost we may become by gliding rapidly between places, however oblivious to place we may be in our thought and theory, and however much we may prefer to think of what happens in a place rather than of the place itself, we are tied to place undetachably and without reprieve" (Casey, 1993, p. xiii).

THE CENTRAL DIMENSIONS OF A SENSE OF PLACE

"(T)he qualities I associate with a sense of place: a lively awareness of the familiar environment, a ritual repetition, a sense of fellowship based on shared experience," writes John Brinckerhoff Jackson (Jackson, 1994, p. 159). What else emerges from these writings the central dimensions of a sense of place? I would suggest at least four propositions for further consideration.

Place has a Geographic or Territorial Dimension.
In many of the formulations cited above, the topographic nature of place emerges as a central feature. For Anderson & Gale (1992), it has an "inherently geographic" nature. For Jackson and Hummon, this dimension reflects the interaction of an individual with a particular environment. This geographic dimension roots a sense of place in a particular location, rendering it specific and endowed with the experiences and rituals associated with that location.

This topographic dimension not only gives the place a specific character. It also means that the place is characterized by particular features. These "fixed" features may be central, but they are hardly unchangeable. In naturalist's descriptions, for example, mountains or oceans may predominate in a description of a place, but it is clear that the landscape will change by time of day, weather and season. Thus the constant features of place are not static but may remain mutable across time and changing conditions.

Places have Boundaries.
Throughout his studies of place, Edward Casey repeatedly illustrates the dimensions of place by contrasting it with its opposite, "placelessness." To be placeless is to be, quite literally, adrift at sea; such a location, without boundaries

and the means to orient oneself in geographic and temporal space, defies the very idea of what he means by place. To be in a place, to know it, means to limn its boundaries. Like the geographic dimension, this characteristic gives place a rootedness and particularity. As Leach writes:

"To be sure, the boundaries of place may, in the nature of things, 'exclude the outsider and the stranger,' " as J. B. Jackson once observed in *The Necessity for Ruins*. But as Jackson also observed, boundaries "stand for law and permanence," "create neighbors," and "transform an amorphous environment into a human landscape." (Leach, 1999, p. 179–180).

Boundaries make the place specific, but they do more than that. They mark the known. They remind us that like community, which by definition implies some criterion for membership, an important part of a sense of place is being able to distinguish where the place begins and ends.

Place is Embued with Social and Sultural Meanings.
Almost all of the definitions above emphasize that a sense of place is socially constructed. Places embody our culture; as Jackson notes, there may be even a ritual element to them. As Casey (1993) suggests:

> For the most part, we get into places together. We partake of places in common – and reshape them in common. The culture that characterizes and shapes a given place is a shared culture, not merely superimposed upon that place but part of its very facticity. Place as we experience it is not altogether natural. If it were, it could not play the animating, decisive role it plays in our collective lives. Place, already cultural as experienced, insinuates itself into a collectivity, altering as well as constituting that collectivity. Place becomes social because it is already cultural (Casey, 1993, p. 31).

Obviously the cultural meanings of place vary according to the groups that enact those meanings. Many places are characterized by social interaction and cultural activities of more than one group of people. The sense of place is only as varied as the groups who make meaning there, and is inclusive only to the degree that the experiences of all participants in those activities are respected and reflected. As indicated earlier, attention to this dimension of place may show that the social experiences of all individuals who construct the place may not be valued equally.

Place has a Temporal Quality, Linking it to Both Past and Future.
Finally, many of these formulations suggest that places are interwoven with experience over time. As Casey notes, "It is by the mediation of culture that places gain historical depth (Casey, 1993, pp. 31–32)." Platt (1996) reminds us that this sense of history also points and directs us toward the future, suggesting that "Places of experience provide categories for managing new adventures and

new cycles of old adventures. Places of experience connect the past to the future, memory to expectation, in an invigorating way" (Platt, 1996, p. 112).

We will return to most of these dimensions of place in a final section that explores their implications for re-imagining the neighborhood school. I would turn now however to some arguments about the role that schools can play in developing social capital in their communities and in their cities. This formation of social capital is often connected to deep and particular understandings about neighborhoods and communities.

SCHOOLS AND THE DEVELOPMENT OF SOCIAL CAPITAL

The concept of social capital became widely known in educational circles with the publication of a study by James Coleman and Thomas Hoffer in 1987. This research compared achievement across public and private high schools, using the 'High School and Beyond' data collected from a nationally representative sample of high schools (Coleman & Hoffer, 1987). The authors argued that students who had access to more "social capital" did better in schools than those who did not because they enjoyed a broad system of support in their communities which helped to foster their school achievement. Social capital was found in the overlapping web of relationships in the communities in which they resided, a network that resulted in high levels of communication, trust and assistance. In later work Coleman defined the concept as ". . . not a single entity, but a variety of different entities having two characteristics in common: They all consist of some aspect of a social structure, and they facilitate certain actions of individuals who are within the structure" (Coleman, 1990, p. 302). The essential characteristic of social capital, in other words, is the fact that it resides in the relationships among individuals within some sort of social organization. The educational benefits accrue when the community as a whole values education and shares some degree of oversight for all children.

In the last two decades the construct of social capital has been used to explain the differences among schools and to investigate ways in which schools can understand and support children and their families. More recently, scholarship has focused on the ways in which schools can become agents in the development of social capital that yields broad benefits for families, communities and cities (Driscoll & Kerchner, 1999).

In that earlier paper, we discussed several different means through which schools help to build social capital in communities (Driscoll & Kerchner, 1999). For instance, we noted that some ways of thinking about the relations between schools and homes emphasize the web of support than can be created by both for all children. Researchers such as Joyce Epstein (1994), for example, delineate a broad framework in which the parent community and the school work together to build these supportive structures. "The term 'school, family and community partnerships' is a better, broader term than 'parent involvement' to express the shared interests, responsibilities, investments, and the overlapping influences of family, school and community for the education and development of the children they share across the school years," she writes (Epstein, 1994, p. 39).

Schools have also moved beyond their relationships with parents and students to become centerpieces in the kind of institutional restructuring that develops social capital in communities. A good example of this approach is the rapid proliferation of coordinated children's services models that often target children at risk. *Within Our Reach*, Lisbeth Schorr's 1988 call to arms for the restructuring and coordination of social services designed to assist children at (Schorr, 1988), argues persuasively that social services must be redesigned in order to avoid "rotten outcomes" such as poor health, poor performance in school, and even imprisonment. Schools help to build the capacity to deal the problems children have with respect to nutrition, medical care, housing conditions, family support, and educational services.

An early example of a program designed to help schools build social capital in their communities is James Comer's School Development Program. (Comer, 1980; Comer, Haynes & Joyner, 1996). The development of capacities for improved educational support in all aspects of children's lives are clearly espoused goals of the program, and the activities are designed to address many facets of development for students, parents and teachers and other adult members of the surrounding community. This program is but one example, however, of the kind of coordinated services efforts inventoried by Louise Adler. Adler notes that most of these restructuring efforts include local access to services at a school or neighborhood institution by families and children; the availability of a variety of services, including heath, mental health, employment, childcare, and education; collaboration among all service providers; a developmental, supportive model and a move toward the empowerment of families and community; flexibility in funding; the development of new ways of working among diverse professionals; and some requirements for change at a systemic level (Adler, 1994, p. 1).

As the movement has developed in breadth and scope, an extensive literature has emerged that documents existing efforts and (more recently) critiques the philosophical and conceptual basis for school linked efforts (see for example, Adler & Gardner, 1994; Haertel & Wang, 1997).

As this volume acknowledges, current thinking on service coordination has moved much more boldly towards a renewed commitment to change in a bigger sense and a rededication to the problems found in the community at large in which the school is located (Crowson, 1998.) As Adler notes, "We cannot get better childhoods for children unless we build better communities" (Adler, 1994, p. 1). The task is formidable, and the capital formation extends beyond the development of utilitarian relationships and improved home-school connections. Asserts Adler: "If we are serious about prevention, we must focus on how to improve the economic viability of communities. Unless we address issues of economic empowerment, we will never be able to build enough homeless shelters, provide enough compensatory education programs, or move more families off welfare rolls than come onto them" (Adler, 1994, p. 10).

Clearly, these efforts to move social service restructuring in the direction of community development may be connected in important ways to knowledge about particular communities. Driscoll and Kerchner (1999) also suggest that there is another powerful way to think about how education can develop social capital in neighborhoods, communities and cities. We postulated that schools are essential tools in the revitalization of cities and communities, and that education is one of the "basic industries" (Kerchner, 1997) that attracts and retains a vital citizen community.

When thinking about education as a tool of community revitalization, it is important to understanding the productive role of neighborhoods and the role of communities as zones of production. "Because occupational groups that cluster in neighborhoods serve to sustain and reproduce the skills, attitudes, and norms necessary to take up an occupation, neighborhoods become places of acculturation and signifiers of social status and labor market capacity" (Driscoll & Kerchner, 1999, p. 395).

This view means that neighborhoods produce and transmit knowledge about how to access this economic activity and can provide role models of adults engaged in these endeavors. Neighborhood schools, then, have an important role to play in seeing that these stocks of social capital are transmitted throughout the community and that all students have access to them. "Social capital flows are not simply tied to increases in measured cognitive achievement; they have profound developmental effects. They reproduce skills and expectations. They can also contribute to the creation of a new dynamic, lead to a change in expectations, and refresh ideas about civic engagement. Schools are among the most stable and important institutions we have for sustaining and transforming these flows of social capital" (Driscoll & Kerchner, 1999, p. 396).

We also argued that education is itself a basic industry that works to keep cities vital in at least two ways. One mechanism construes schools as a magnet that attracts some individuals into a community and helps others to stay. The advantages of a strong school system among affluent middle class families is known well by both big city mayors and real estate agents. This clientele provides an important taxpayer constituency as well as political support. Keeping schools strong is an important tactic in attracting other industries, a fact that is sometimes forgotten when inducements such as tax abatements are offered that can actually diminish schools' abilities to raise adequate funds fairly (Driscoll & Kerchner, 1999, p. 398).

A second mechanism imagines schools as engines, that is, as a major employer in the community with the power to direct resource flows to the community's benefit. As McKnight and others have argued, schools can engage in such practices as purchasing goods and services from local producers and suppliers; hiring local residents; targeting contracts for goods and services to support the creation of new businesses; and investing resources in local financial institutions such as credit unions, co-ops and community development loan funds (McKnight, 1994, 1995; McKnight & Kretzmann, 1993).

This latter mechanism in particular connects schools with the places in which they are located and sees educational institutions as vital and critical players in the lives of those places.

Schools also serve as "training tracks for democracy," providing a civic focus in some communities. Schools provide an opportunity for students to enact citizen roles and function as the place for public discourse and debate about purpose and practice (Driscoll & Kerchner, 1999, p. 399). By serving as public spaces, both literally and in an intellectual sense, educational institutions can be the very bedrock of community, a place where a socially constructed narrative of growth and learning is enacted on a daily basis.

DEVELOPING SOCIAL CAPITAL WITH AND WITHOUT A SENSE OF PLACE

The idea of a sense of place can be a useful lens through which to examine some of the effects of these attempts to develop social capital through education. For example, research on coordinated children's services described above (Smrekar & Mawhinney, 1999; Driscoll, Boyd & Crowson, 1996) has documented the particularly difficult nature of inter-professional relationships in these school-based programs. Far more common than the collaboration originally envisioned is a mere co-location of services in which shared language and perspectives fail to develop. In part this is the result of structural arrangements that allow little time for prolonged interaction or deep discussions among professionals working with clients.

But I would argue that it precisely the lack of a sense of place in many of these coordinated service arrangements that makes them difficult. While patterns of interaction and behavior may vary from project to project, the very nature of the "collaboration" is rooted in bringing together standardized professional communities in service of clients who are not part of these same discourses. Many of the collaborations are not rooted in a shared understanding of a particular community's needs, but rather spend their time negotiating the often uneasy coexistence of the norms of the collaborative agents such as social workers, teachers, health professionals and others. Thus these efforts may never take on the shape and identity of the specific communities they seek to serve. The cultural and historical understandings of the community may weigh little in comparison to the professional norms of service that drive these efforts.

John McKnight's lament in his essay "John Deere and the Bereavement Counselor" seems particularly apt given this problem. When service professionals supplant those functions that were once part of the community's care for itself, he argues, their benefits must be "weighed against the sum of socially distorting monetary costs to the commonwealth; the inverse effects of the interventions; the loss of knowledge, tools and skills, regarding other ways; and the antidemocratic consciousness created by a nation of clients" (McKnight, 1996, p. 10). Neglecting the needs of specific communities and supplanting their ability to care for themselves in favor of more universalistic, professionally driven relief efforts, in other words, carries a dear price. The challenge is to provide for the community in ways that the community can recognize as worthwhile and sustain as needed.

Another reform that builds social capital, based on the work of McKnight and others, is much more in tune with the idea of a sense of place. Kretzmann & McKnight (1993, 1996) describe a model of community development that they term "asset based." In this model, communities are thought of as unique collections of resources and assets, and activities are designed to advance the conceptual and physical mapping of communities. This model is intended to be an alternative to images of communities, especially in economically challenged areas, as collections of deficits or needs. While asset based community development does not rule out securing resources that are external to the community, it begins with the development of an inventory of the community's assets generated from domicile to domicile. It also imagines the ways in which institutions located in the community can be "reclaimed" in service of development.

Although this model is not specifically educational in focus, it recognizes the importance and the central roles of schools as institutions in which many vital assets are collected. As such, the model is quite directive with respect to the ways in which community developers should connect with, and eventually reconfigure, their schools. Kretzmann and McKnight outline a multidimensional map that inventories the resources of the school and connects these with the previously inventoried assets of the individual members of the school community. The combination of these asset lists results in a "strong, concrete, mutually beneficial partnership" between local schools and community individuals, organizations, and associations. Among the strategies suggested are "capturing" physical and personnel resources, including the school building and staff, as well as the development of financial and commercial relationships between the community and the school that harness the schools purchasing power and its ability to build and secure additional financial capacity.

This type of reform may not be derived from the educational community, but – writ large – clearly has the potential to affect schools and the way they connect to their surroundings. It has many of the hallmarks, I would argue, of a plan that is deeply embedded with a sense of place. The tasks involved are not standardized or neutral but derived from a local and highly detailed assessment of the community and the school itself. Although the "process" of development can be outlined, it runs no sure course and will vary according to the places and the spaces – physical, social, cultural and historical – that comprise its elements.

THE SENSE OF PLACE AND ITS IMPLICATIONS FOR NEIGHBORHOOD SCHOOLS

Education has a role, then, in the development of the social capital that supports communities and keeps neighborhoods healthy and vital. Neighborhood schools can be centers of community activity and development. In conclusion, I want to outline some of the implications of a sense of place for understanding what I will call the "re-imagined" neighborhood school. These implications include:

(1) a renewed appreciation of the physical settings of the school and its importance in the neighborhood;
(2) a commitment to thinking about issues of safety and violence in a way that construes the borders of the educational place as extending beyond the school building;
(3) a deep awareness of the importance of history in the social construction of neighborhoods and their schools; and, finally,
(4) an understanding of the paradox that it is only when we know and live in an educational place that we may be prepared to leave it.

1. The physical setting of the school requires attention, care, and imagination, and must embody the values and priorities of the neighborhood and community.

"How is it that we have moved from a time when the schools we built were significant, even monumental, to a time when we resist establishing even the most minimum standards of quality for our public schools?" laments Cynthia Uline. "In so many, we continue to justify the violation of basic health and safety codes, because the doors must be ready to open each Monday morning. We tell ourselves that teachers' simple classrooms should suffice, lead-based paint, asbestos and overcrowdedness notwithstanding. We have allowed our schools to fall into a state of disrepair we would not accept in the places where we work, eat, recreate or shop" (Uline, 2000, p. 446).

She continues:

> Many maintain that the only concerns and solutions are of a fiscal nature. In discussions of capital outlay and facilities planning, we must stick to tough-minded concerns while ideas of comfort, beauty and imagination are regarded as esoteric, pretentious, and expensive. Still, if we wish our schools to support community values and be centers of excellence, perhaps it is time to see a more thoughtful balance. In minimizing the importance of what might be termed internal concerns, perhaps we are denying communities and invigorating and engaging opportunity (Uline, 2000, pp. 446–447).

The central point here is, as Uline asserts, that "A school building sends a message to its occupants and to the community beyond" (Uline, 2000, p. 449). Citing research that documents the educational benefits of well designed, pleasing schools, Uline argues that our understandings of education are too often subjugated to the cost-driven concerns of the facilities planners we deem the experts. Not so, she reminds us, with early public educators such as Henry Barnard, whom she calls a

> prominent example of one educator who combined school design and pedagogy, recognizing the importance of community engagement and commitment . . . Barnard extends his idea of school beyond its physical place, seating it within the character of community and the nurturing of its commitment to education as a shared interest" (Uline, 2000, p. 456).

Understanding what a sense of place means in educational terms, both for the community and the teachers and learners themselves, is a critical task that too often becomes bureaucratized and relegated to outsiders. Uline argues that the body of research that begins to address issues of place is separated from the mainstream knowledge of teachers and administrators. She suggests that "(w)hen it comes to questions about teaching and learning and administering, perhaps the mind needs to think about place to inform itself about the activities that will occur within that place. Certainly this was the case with Barnard, for whom place, pedagogy, and community character were inextricable" (Uline, 2000, p. 456–457).

We are reminded, then, that the quality of the spaces we construct and maintain for educational purposes matter. They cannot be accidental. They can not be badly maintained and still be uplifting or illuminating. They cannot neglect the social functions that must take place within them nor forget the need to create public spaces where community life can be enacted.

2. The boundaries of the places we call school must be constructed in ways that respect the neighborhood and construe the school and its community as an integrative whole that supports students.

If boundaries are inherent to a sense of place, we must be careful where we draw them. The physical structure of the school itself has often been the delimiting line. In this view, the community is construed as something akin to the three-mile limit – a place that stands between the safe shore and the vast unknown ocean beyond, mapped to some degree but still a potential source of danger.

Safety and security in and around the school are paramount concerns of educators and public alike, even without recent media attention to a series of incidents of school violence in otherwise "quiet" communities. A recent national survey indicates that parents worry more about crime than they do the quality of public education in both urban and suburban communities (Carnevale & Desrochers, 1999). But it is an oxymoron to say that we can sustain safe schools in unsafe communities; it is worse to believe that the community itself does not share primary concern for school safety.

A sense of place for a neighborhood school must embrace a sense of security that moves away from a "fortress" mentality. Security comes from knowing who we are, not in keeping the community out. A broader imagination of a neighborhood school considers how the social capital in the community itself helps to make children safe, and draws on the relational caring that supports children through broad networks of engagement and oversight.

3. The historical dimensions of place should animate our neighborhoods and root the educational conversations in ways that are productive towards the future. We must explore the many histories and cultures of a place rather than impose just one.

"No place is a place until things that have happened in it are remembered in history, ballads, yarns, legends, or monuments," says Wallace Stegner. He continues:

Fictions serve as well as facts. Rip Van Winkle, though a fiction, enriches the Catskills. Real-life Mississippi spreads across unmarked boundaries into Yoknapatawpha County. Every one of the six hundred rocks from which the Indian maiden jumped to escape her pursuers grows by the legend, and people's lives get lived around and into it. It attracts family picnics and lovers' trysts. There are names carved in trees there. Just as surely as do the quiet meadows and stone walls of Gettysburg, or the grassy hillside above the Little Big Horn where the Seventh Cavalry died, even a "phony" place like the Indian maiden's rock grows by human association (Stegner, 1992, p. 202).

We are used to thinking about the ways in which schools can connect to the cultural narrative of a community. For most of the last century, the names of new schools often commemorated the ethnic heroes of the community. Naming a school for the Irish patriot Robert Emmett, for example, was a signal to the Irish families who settled in Chicago that their history was not forgotten, just as naming a school for the ballplayer Roberto Clemente commemorated his accomplishments to the Puerto Rican community some seventy years later.

But the heroes we remember need not speak to one cultural group in the community only. We are all greatly enriched when the narrative of a neighborhood or another culture becomes part of our own. As communities have shifted and new immigrants have replaced the old in these neighborhoods, the significance of the school place name may get lost. Wallace Stegner reminds us that "Communities lose their memories along with their character. For some, the memory over time can be reinstated" (Stegner, 1992, p. 203).

Our connection to history – our own, and to that of others' – may be unfamiliar, but is still a critical part of the development of a community. As Stegner observes: "History was part of the baggage we threw overboard when we launched ourselves into the New World. We threw it away because it recalled old tyrannies, old limitations, galling obligations, bloody memories" (Stegner, 1992, p. 206). But we jettison that historical knowledge at our peril, he reminds us: "Neither the country nor the society we built out of it can be healthy until we stop raiding and running, and learn to be quiet part of the time, and acquire the sense not of ownership but of belonging" (Stegner, 1992, p. 206).

People's lives get lived around and into the places we call schools, and until we embrace the historical elements of our schools and our communities we will never feel the sense of belonging that these places can convey. With these histories – not merely the larger artifacts of the group's ethnic cultures, but also the specific place history that has been lived as the saga of the school itself – come challenges, and, as Stegner reminds us, recollections of things we may wish to forget. But only by acknowledging from whence we have come can we begin to plan an accurate road map for the future.

4. Places and neighborhood schools are socially constructed over time. Precisely because we expect most people to leave the neighborhood we must make sure they are rooted long enough to know it.

"If you don't know where you are, says Wendell Berry, you don't know who you are," begins Wallace Stegner's essay "The Sense of Place" (Stegner, 1992, p. 199). He continues, "He (Berry) is not talking about that kind of location that can be determined by looking at a map or a street sign. He is talking about the kind of knowing that involves senses, the memory, the history of a family or a tribe. He is talking about the knowledge of place that comes from working it in all weathers, making a living from it, suffering its catastrophes, loving its mornings or evenings or hot noons, valuing it for the profound investment of labor and feeling that you, your parents and grandparents, your all-but-unknown ancestors have put into it" (Stegner, 1992, p. 205).

The commitment to work a place in all weathers – to nurture its development over time, whether it be a neighborhood or a school that serves as the linchpin of that community – is perhaps the essential ingredient in the creation of a sense of place. For a sense of place rests on repeated activities, the social construction of shared knowledge over time, prolonged interaction with an environment, and knowledge of as well as participation in the history created there.

Stegner's essay, much like Proefriedt's reflections, discusses the American restlessness that makes so many of us "displaced" persons. The romantic lure of the unknown and the promise of something better – like Wolfe's train to the big city – are in part to blame. As he writes:

"Indifferent to, or contemptuous of, or afraid to commit ourselves to, our physical and social surroundings, always hopeful of something better, hooked on change, a lot of us have never stayed in one place long enough to learn it, or have learned it only to leave it." That restlessness comes at a price, Stegner concludes. "In our displaced condition we are not unlike the mythless man that Carl Jung wrote about, who lives 'like one uprooted, having no true link either with the past, or with the ancestral life which continues within him, or yet with contemporary human society. He . . . lives a life his own, sunk in a subjective mania of his own devising, which he believes to be the newly discovered truth" (Stegner, 1992, pp. 204–205).

Henry Adams traveled the world throughout most of his lifetime; indeed, the descriptions of Boston and Quincy comprise but a small portion of his narrative. But it was to Boston and Quincy he returned in his memory when he first set out to make sense of those travels, and to understand the path of his life. That sense of place rooted him, and entwined as it was with his recollections of his educational experiences, it provided a map that guided him well in the years to come. I would suggest that the neighborhood school of this next century must play the same role, and create for our children a sense of place that links them to a deep understanding of their communities as the best preparation for venturing into the wider world beyond.

REFERENCES

Adams, H. (1918/1961). *The education of Henry Adams*. Boston: Houghton Mifflin Company.

Adler, L. (1994). Introduction. In: L. S. Adler & S. Gardner (Eds), *The Politics of Linking Schools and Social Services* (pp. 1–16). Washington, D.C.: The Falmer Press.

Adler, L., & Gardner, S. (1994). *The politics of linking schools and social services*. Washington, D.C. & London: The Falmer Press.

Altman, I., & Zube, E. (1989). *Public places and spaces*. New York: Plenum Press.

Anderson, K., & Gale, F. (1992). Introduction. In: K. Anderson., & F. Gale. (Eds), *Inventing Places: Studies in Cultural Geography* (pp. 1–14). Melbourne, Australia: Longman Cheshire.

Capper, C. (1994). We're not housed in an institution, we're housed in the community: Possibilities and consequences of neighborhood-based interagency collaboration. *Educational Administration Quarterly, 30*(3), 257–277.

Carnevale, A., & Desrochers, D. (1999). *School satisfaction: A statistical profile of cities and suburbs*. Princeton, N.J.: Educational Testing Service.

Casey, E. (1993). *Getting back into place: Toward a renewed understanding of the place-world*. Bloomington, IN: Indiana University Press.

Chaskin, R. J., & Richman, H. (1992). Concerns about school-linked services: Institution-based versus community-based models. *Future of Children, 2*(1), 107–117.

Conant, J. B. (1959). *The American high school today*. New York: McGraw Hill.

Coleman, J. (1990). *Foundations of Social Theory*. Cambridge, MA and London, England: The Belknap Press of Harvard University Press.

Coleman, J., & Hoffer, T. (1987). *Public and private schools: The impact of communities*. New York: Basic Books.

Comer, J. (1980). *School power: Implications for an intervention project*. New York: The Free Press.

Comer, J., Haynes, N., & Joyner, E. (1996). The school development program. In: J. Comer, N. Haynes, E. Joyner & M. Ben-Avie (Eds), *Rallying the Whole Village: The Comer Process for Reforming Education* (pp. 1–26). New York: Teachers College Press.

Crowson, R. (1998). Community Empowerment and the Public Schools: Can Educational Professionalism Survive? *Peabody Journal of Education, 73*, 56–68.

Crowson, R., & Boyd, W. (1993). Coordinated services for children: Designing arks for storms and seas unknown. *American Journal of Education, 101*(2), 140–179.

Driscoll, M. E., & Kerchner, C. (1999). The implications of social capital for schools, communities and cities: Educational administration as if a sense of place mattered. In: J. Murphy & K. Louis (Eds), *Handbook of Research on Educational Administration* (2nd ed.) (pp. 385–404). San Francisco: Jossey Bass.

Epstein, J. L. (1992). School and family partnerships. In: M. Alkin, (Ed.), *Encyclopedia of Educational Research* (pp. 1139–1151). New York: Macmillan.

Epstein, J. L. (1994). Theory to practice: School and family partnerships lead to school improvement and student success. In: C. Fagnano & B. Werber, (Eds), *School, Family and Community Interaction: A View from the Firing Lines* (pp. 39–54). Boulder, Co: Westview Press.

Hiss, T. (1996). *The experience of place.* New York: Vintage Books.

Hummon, D. (1992). Community attachment: Local sentiment and a sense of place. In: I. Altman & S. Low (Eds), *Place Attachment* (pp. 253–278). New York: Plenum Press.

Hummon, D. (1990). *Commonplaces: Community ideology and identity in American culture.* Albany, NY: SUNY Press.

Jackson, J. B. (1994). *A sense of place, a sense of time.* New Haven, CT: Yale University Press.

Kerchner, C. T. (1997). Education as a City's Basic Industry. *Education and Urban Society, 29*(4), 424–441.

Kretzmann, J., & McKnight, J. (1996). Assets-based community development. *National Civic Review, 85,* 23–29.

Kretzmann, J., & McKnight, J. (1993). *Building communities from the inside out: A path toward finding and mobilizing a community's assets.* Evanston, IL: Asset Based Community Development Institute of the Institute for Policy Research at Northwestern University.

Lasch, C. (1990). *The true and only heaven: Progress and its critics.* New York: Norton.

Leach, W. (1999). *Country of exiles: The destruction of place in American life.* New York: Pantheon.

McKnight, J. (1995). *The careless society: Community and its counterfeits.* New York: Basic Books.

McKnight, J. L. (1994). Hospitals and Communities Create Wise Environments. *Trustee, 47*(2), 22–23.

McKnight, J. L., & Kretzmann, J. (1993). *Building Communities From the Inside Out.* Evanston, IL: Northwestern University.

Monk, J. (1992). Gender in the landscape: Expressions of power and meaning. In: K. Anderson & F. Gale. *Inventing Places: Studies in Cultural Geography* (pp. 122–138). Melbourne, Australia: Longman Chesire.

Platt, K. (1996). Places of experience and the experience of place. In: L. Rouner (Ed.), *The Longing for Home* (pp. 112–127). Notre Dame, IN: University of Notre Dame Press.

Proefriedt, W. (1985). Education and moral purpose: The dream recovered. *Teachers College Record, 86*(3), 399–410.

Smrekar, C., & Mawhinney, H. (1999). Integrated services: Challenges in linking schools, families and communities. In: J. Murphy & K. Louis (Eds), *Handbook of Research on Educational Administration* (2nd ed.) (pp. 443–462). San Francisco: Jossey Bass.

Smylie, M., Crowson, R., Chou, V., & Levin, R. (1994). The principal and community-school connections in Chicago's radical reform. *Educational Administration Quarterly 30,* 342–364.

Spain, D. (1992). *Gendered spaces.* Chapel Hill, NC: University of North Carolina Press.

Steele, F. (1981). *The sense of place.* Boston: CBI Publishing.

Stegner, W. (1992). The sense of place. In: W. Stegner (Ed.), *Where the Bluebird Sings to the Lemonade Springs: Living and Writing in the West* (pp. 199–206). New York: Penguin Books.

Tyack, D. (1974). *The one best system.* Cambridge: Harvard University Press.

Tyack, D., & Hansot, E. (1982). *Managers of Virtue: Public School Leadership in America, 1820–1980.* New York: Basic Books, Inc.

Uline, C. L. (2000). Decent facilities and learning: Thirman A. Milner Elementary School and beyond. *Teachers College Record, 102*(2), 442–460.

Winchester, H. The construction and deconstruction of women's roles in the urban landscape. In: K. Anderson & F. Gale, *Inventing Places: Studies in Cultural Geography* (pp. 139–156). Melbourne, Australia: Longman Chesire.

Wolfe, T. (1934/1968). *You can't go home again.* New York: Harper and Row.

LEADERSHIP OUTSIDE THE TRIANGLE: THE CHALLENGES OF SCHOOL ADMINISTRATION IN HIGHLY POROUS SYSTEMS

Charles Taylor Kerchner and Grant McMurran

INTRODUCTION

For most of the past century students of organizations have occupied themselves poking holes in the iron triangle of hierarchy. Penetrating the walls of hierarchy moved organizational thought from closed systems to the general acceptance of open systems. Contemporary conceptions of organizations as networks further challenge our view of organizations and how one leads them. Old ideas about efficiency and control fail to explain how networked organizations work (Clegg, 1990).

By any reasonable standard, public schools have become highly porous hierarchies. The current era of educational reform consists almost entirely of those outside the hierarchy attempting to change those inside. More importantly, it is now well recognized that vital resources to carry out schools' core functions must be imported. Ultimately school success rests on activating personal, family, and community resources. Our perception of schooling has moved a long way from Cubberly's (1916) depiction of children as objects of the state to the contemporary recognition that schools control perhaps half of what accounts for measured cognitive achievement. In an era in which schools are publicly

Community Development and School Reform, Volume 5, pages 43–64.
ISBN: 0-7623-0779-X

graded and ranked according to their achievement outputs, school leadership faces a substantive crisis of control. How can one lead when one's place in the hierarchy does not provide either the essential resources or the authority over production of core results?

We believe the answer to this leadership question lies in the creation and activation of social capital. As a theoretical concept, social capital links the micro and institutional levels in society allowing a causal argument to be made. In everyday practice, creating social capital formation allows school administrators to access motivation and control resources they need and don't possess inside the organization. In pursuit of insight on this question, we examine one Southern California school district, Pomona Unified, and its role in the formation and activation of social capital.

SOCIAL CAPITAL AND POMONA

The city of Pomona rests at the eastern end of Los Angeles, west of the Ontario International Airport and southwest of the Claremont Colleges. Incorporated in 1888, Pomona started out as a sleepy farming town that was transformed into the commercial center of eastern Los Angeles County by the railroad. During and after World War II the city was incorporated in the web of aerospace manufacturing that proliferated in Southern California.

Starting in the early 1960s Pomona underwent major economic and demographic shocks. With the growth of the metropolitan area and the completion of a freeway, Pomona became increasingly suburban. While the defense cutbacks of the 1990s devastated Southern Californian industry, Pomona felt the loss of its industrial base years earlier. Brand-name retail also vacated the city, first to shopping centers on the edge of town, and then to larger regional malls. Today the city is left with a classic but underutilized downtown and a partly abandoned shopping center, both reminders that Pomona was once a place that everyone went to.

During the 1960s Pomona experienced a rapid growth of African American residents. Twenty years later, it became a settlement community for Latinos, many recently arrived in the United States. By 1990, 51% of the city's 140,000 residents were Hispanic while another 13% were African American. Thus, in 30 years, the Pomona Unified School District was transformed twice, from a district serving mostly middle class Anglo children, to a one with a large African-American student body, to one in which Hispanic children are the majority.

Bordered by considerably more affluent cities to the north, west and south Pomona stands out as a relatively low-income community. Neighboring cities of Claremont, La Verne, Covina, and Diamond Bar all have far greater median household incomes than in Pomona (see Map 1).[1] This income disparity is also reflected in lower housing values and a relatively large number of home renters. Over 40% of Pomona residents do not own their own homes. A generation ago, Pomona stood out in the region as a center of commerce; today it is an example of urban decline.

The Pomona Unified School District serves most of the municipality of Pomona including all of its central city area. Although dwarfed by the Los Angeles Unified School District, it is at 32,789 students (1998) among the state's largest school agencies. The district operates 27 elementary schools, six middle schools, five high schools and an alternative school

The Social Capital Concept

In his 1990 treatise *Foundations of Social Theory*, James Coleman presents a detailed exposition of the theory of social capital as a construct that can help amplify classical understandings of rational behavior. Coleman credits Loury (1977, 1987) with the introduction of the term 'social capital' (Coleman, 1990, p. 300). He cites as well the work of Bourdieu (1980) and Flap and De Graaf (1986), all of whom, he suggests, have used similar terms (Coleman, 1990, p. 300).

As a concept, social capital is situated within the broader theoretical context called "new institutional economics" Granovetter (1985). This contemporary economic theory rebuts what Coleman calls "broadly perpetuated fiction in modern society," that "society consists of a set of independent individuals, each of whom acts to achieve goals that are independently arrived at, and the functioning of the social system consists of the combination of these actions of independent individuals" (Coleman, 1990, p. 302). (For a more complete introduction to the concept of social capital and its implications for educational leadership, see Driscoll & Kerchner, 1999.)

Pomona and Leadership Through Social Capital

The Pomona Unified School District has all the leadership challenges associated with declining central cities. Over the past quarter century its governance has been often characterized by a quarrelsome school board that gave incumbent superintendents razor-thin majorities and generally a short term in office. As a result many of its schools, particularly the secondary schools, took on the characteristic of fiefdoms that offered ritual loyalty to the district but which operated with great independence.

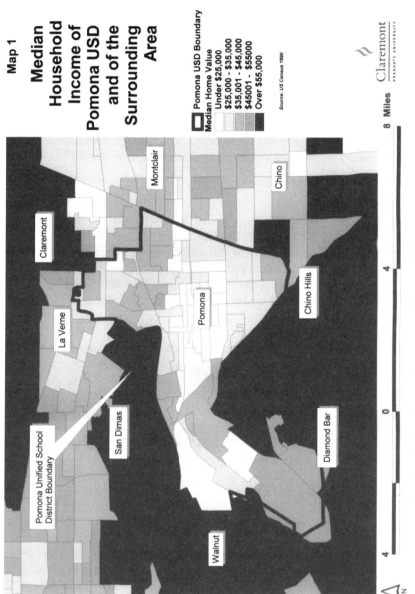

Map. 1. Median Household income of Pomona USD and the surrounding area.

Along with the problems of stability and control, the district began to realize that much of what it needed to be successful could not be created through yet another categorical program or special service. Success in schooling began to be associated with success for families and generally for the community. Attacking the problem in Pomona began by realizing that family resources that had been thought of as personal and private were, in effect, part of the public resources that made schools work. How was it that the private resources of poor and working class families, many of them recent immigrants, could sway the balance of school effectiveness?

The answer, of course, was that in the economic sense, family resources are not exclusively private resources. When they flow, they have enormous public effect. The "public-good aspect of social capital" explains at least in part the fact that social capital is underestimated, understudied, and for the most part overlooked. In essence, "many of the actions that bring social capital into being are experiences by persons other than the person so acting. The result is that most forms of social capital are created or destroyed as a byproduct of other activities. Much social capital arises and disappears without anyone's willing it in or out of being; such capital is therefore even less recognized and taken into account in social research than its intangible character might warrant" (Coleman, 1990, pp. 317–318).

In other words, it is *most* often the case that people who do not directly create social capital benefit from its existence. This is the case whether participation takes the form of an active parent-teacher association that continues to provide benefits for the school even though not all parents participate, or the protection of a neighborhood from theft because the Neighborhood Watch Program provided by a few residents has succeeded in reducing the number of thieves.

Robert Putnam's study of civic traditions in modern Italy (Putnam, 1993) also emphasizes the ways in which social capital is a public good. He postulates that there are few losers in groups characterized by uniformly high levels of activity or investment in the networks and organizations that promote social capital:

> Stock of social capital, such as trust, norms, and networks, tend to be self-reinforcing and cumulative. Virtuous circles result in social equilibrium with high levels of cooperation, trust, reciprocity, civic engagements, and collective well-being.

Moreover, these networks "facilitate communication and improve the flow of information about trustworthiness of individuals;" likewise, the networks make it easier for participants to cooperate and "embody past success at collaboration" (Putnam, 1993, p. 174). For Pomona, the question was how to create flows of civic contribution.

The Village @ Indian Hill: Pomona's New Venture

The Village @ Indian Hill was a dying commercial mall at the intersection of what had been two major shopping streets. Since the major anchor stores left to a larger mall located nearby, the mall experienced an exodus of retail that left only a small number of low-end stores. As a consequence, much of the 700,000 square foot facility was either unused or underused.

The demise of retail at the mall provided the Pomona Unified School District with multiple possibilities for development. Looking to overcome its acute facility shortage, the Pomona Unified School District invested in developing the Village. Since the space at the mall could be purchased or leased cheaply, the district found that the existing structure could accommodate students more economically than by building new schools or even leasing portable classrooms. The school district established an elementary school within the mall serving over 800 students. It is constructing four additional schools with associated teacher professional development facilities. Eventually, it would like to turn these spaces into academy-type high schools.

Administration facility shortage was also overcome by using excess space at the mall. The district's information resources department moved into the Village along with its child development and learning resources departments. The school district's investment has stimulated other growth as other organizations are also investing in the Village such as a NASA/JPL applied technology science classroom, a training center, an adult education center, and a visual arts and performing art center.

The district's involvement in the Village @ Indian Hill has been found to be a sound real estate investment. The rebound of Southern California real estate prices has increased land values within Pomona. Vacant stores that once surrounded the mall are starting to once again receive tenants. The ugliness and blight associated with urban decay appears to have gotten better around the Village.

The late-1990s resurgence of the Southern Californian economy must receive credit for at least part of this revival, yet the investments of the school district should not be overlooked. Converting empty shops to classrooms, district offices, and other community support services brings in potential customers to the stores that remained in the mall. The district took a leading role in cleaning up the mall and making it more inviting for shoppers. By redeveloping the mall, the Pomona Unified School District took a leadership role in revitalizing the city, that in turn increase the value of the district's investment.

Data and Data Collection

Our involvement in the Pomona Unified School District began with a project investigating community perceptions towards the Village @ Indian Hill. The Village has multiple possibilities to the serve the community, but the old shopping center site is clearly problem ridden. Although great efforts were undertaken to improve the image of the Village, many residents have reservations about going there for shopping or other social activities.

Our first task was to gather and construct a demographic picture of Pomona using Geographic Information System (GIS), which allows the display of social statistics and other data on easy-to-read maps. Our GIS system allowed us to incorporate geographic information and combine it with standard demographic information and social statistics. These data included demographics from the 1990 U.S. Census and 1998 updates obtained from a private marketing company. We also used GIS software to examine focus group and survey responses. Using this combination of data with GIS we were able to obtain valuable insights to the strengths and weaknesses of the Pomona area.

Our next task was constructing a survey instrument and a set of focus group questions for parents and residents. We applied a novel GIS techniques developed in a study of Hennepin County, Minnesota (Fulton et al., 1997, pp. 27–28). Generally, the surveys asked resident's opinions about life in Pomona and the focus groups were built around identification of geographic areas that are perceived as community assets and liabilities. Each respondent was given a street map of their neighborhood area and asked to color in areas they liked and disliked, places they felt safe and unsafe, places they shopped or visited. Then the focus groups discussed the places that they marked.

The respondent's maps were then entered digitally into the computer with the help of the GIS software, ArcView. Once in a GIS, the respondent's data were aggregated into composite scores to show areas where many respondents perceived as assets or liabilities. (This representation is shown as Map 3 and discussed later in the chapter.) The GIS system also allowed comparison of resident's perception with other geographic statistics such as police crime data, household income, and land use.

Social Capital and the Multiple Economies of Pomona

Early on in our work it became apparent that success in social capital development in Pomona required rethinking the relationship between the school and community. For the last 80 years, school governance in Pomona has been separate from municipal governance. Throughout the nation, Progressive Era reformers deliberately separated the governance of schools from cities, and they shunned more socially integrated forms of education, such as settlement houses or community centers (Murphy, 1990; Tyack & Hansot, 1982). Educators also embraced theories that enabled them to separate schools from government conceptually. Early human capital economists predicted large public returns to education and made it possible to view schools as inherently *developmental* (Becker, 1962; Schultz, 1981). Organizationalists offered closed-system efficiency theories of functional specialization that led Progressives to professionalized education and separate its functions from those of families and governments.

By the 1970s, however, it was clear that the developmental assumption had broken down. Policy makers and analysts increasingly focused on socially and economically redistributive activities of big city school systems. Redistributive programs, from desegregation aid to Title 1, were expected to be transitory but surburbanization and deindustrialization overwhelmed the effects of such interventions, leaving central cities increasingly divided between rich and poor (Ehrenreich, 1990).

Not surprisingly, the dominant theories of education and society shifted to match urban reality. Human capital theory moved from its assumptions that more education would automatically translate into prosperity toward something close to cynicism. Economic stratification and reproduction theories seemed a much better explanation of urban development, and these rose to prominence among educational sociologists (Rubinson & Browne, 1994). Even the current organization of educational scholarship reflects this shift. Virtually the entire history of the subfield known as the politics of education is devoted to the politics of distribution: equity, voice, agency and interest group politics.

We believe we are now at a turning point in both theory and in the organization of city schools in the United States. There is a thirst for open-system and ecological ideas that show how social and economic subsystems interact to produce both growing metropolitan economies and vibrant, livable neighborhoods. Both at the levels of social theory and of education policy, there is a recognition that schools build as well as benefit from good neighborhoods. But how education plays out its role is a much more complex matter than commonly thought. For much of this century it was thought that any investment in education paid handsome dividends. Educated people made more money. Clusters of educated people formed stronger communities, regions, and nations. Much of the support for universal public education in the United States was based on this assumption, and the assumption is still mostly right. Throughout the globe, places that have invested in education have done better than those which have not.

However, the most recent investigations into the relationship of education and the economy underscore the importance of *how* resources are used to create *linkage* between one sector of a community and others. A community asset is only valuable if it is put to work; a natural comparative advantage is only realized if it is organized around. Flows of resources count rather than stocks.

As we worked in Pomona, we began to think in terms of four economies and the need to find solutions that to the extent possible are cooperative rather than competitive. Otherwise, community politics devolve into non-productive squabbling. We saw:

- the family economy;
- the business economy;
- the municipal economy;
- the school district economy.

Each of these four economies is in some ways competitive with the others. Money a family pays in taxes decreases its spendable income. Diversions from municipal police budgets increase the security costs of schools.

School buildings have this characteristic, too. Almost everyone agrees that they are necessary. A simple glance at the crowded conditions on many campuses is quite convincing. Paying for school buildings is another matter. All means of raising construction revenues have disadvantages; all involve drains on the four economies.

But interdependency of the four economies also creates an opportunity. If investments in each of the economies are made strategically, all four economies can win. So, it is important to consider how the economies link, because it is in finding projects that link the four economies that cities create growth that can be sustained over many years.

Community development research gave us a clue about where to find projects that fit within the sustainable community growth diamond. Each of the economies develops the whole local economy when it induces change in what Hirschman (1958) calls "backward and forward linkages." When school policies and practices spur families, students, or the student's prior school to increase investment in learning, they engage in backward linkages. When schools cause colleges or employers to value their graduates more highly or invest in their further education, they engage in forward linkage.

SCHOOLS AS MAGNETS

Schools serve as magnets for people and for economic activity. In urban areas this phenomenon is largely seen in the negative: families fleeing city schools.

Our focus groups in Pomona revealed substantial numbers of parents who either planned to move out of town or who would move if they could. Their actions and attitudes illustrate that public policy contains far fewer incentives for families with financial means to stay and rebuild cities. In the choice between "voice and exit," we are paid to flee (Hirschman, 1970).

However, city schools are not incapable of being attractive. For whatever their association with social stratification, it is clear that magnet schools, vocational specialty schools, and academically elite schools are highly attractive to parents and that students do relatively well in them (Gamoran, 1996a, b).

In Pomona, one of the challenges has been to develop schools and neighborhoods simultaneously. The availability of reasonably priced housing – bargains compared to surrounding communities – is beginning to draw people to central Pomona. One of the characteristics of the area we studied was the high quality and good repair of its housing. This is particularly true in Lincoln Park, one of the city's most handsome older neighborhoods, but the rest of the area is also pleasant and generally well kept. Said one focus group member: "What I like about Pomona – or our little pocket – is that you have extraordinarily good buys on houses. Lincoln Park, if you could move those homes to Claremont, you would be selling those houses for $200,000 more."

There is anecdotal evidence that these neighborhoods are beginning to attract younger families in search of housing bargains in what will be an increasingly tight housing market as prices escalate following the recession of the early 1990s:

> We just had a gentleman move into the neighborhood who is a writer for Disney; so he can work out of the house. I ran into him at the supermarket the other day. [He said it was] much easier and quieter than it was in Westwood [an expensive residential district on the west side of Los Angeles]. It was harder for him to get work done there. We should aim at some of those kinds of people who work out of their houses.

Unfortunately we also found relatively high levels of dissatisfaction with public schools in these areas. Particularly, when we talked with people in churches and outside of schools themselves, respondents were critical of the public schools and sought alternatives.

Indeed there is a clear difference in homes prices inside the Pomona Unified School District compared to those in neighboring school districts. For example, in north Pomona the median house falls between $100,000 to $150,000 while across the district line in Claremont or La Verne the median home falls between $150,000–$300,000. The same is true in the south with the cities of Chino and Chino Hills and to west with Walnut and San Dimas.

Typically, a house in Pomona will sell for thousands of dollars less than a similar house just outside the school district. Many of the neighborhoods surrounding Pomona were built during the same period and have similar styles and features, but greatly different asking prices. Houses in Pomona do appear to have a "Pomona discount" when compared to similar houses less than two miles away.

While there are many factors that influence home values, the reputation of a school district is a strong factor when considering buying a home. Considering the high cost of private education, it is often in the best financial interest of a parent to relocate to an area with a preferred school district than to pay tuition for each child. The realtor's mantra – location, location, location – surely takes into account whether a home is located within a perceived good school district.

SCHOOLS AS ENGINES

Schools are big employers with huge budgets. If California's elementary and secondary school system were placed on the *Fortune* 500 list of the largest companies in the United States, it would rank 22nd, just above Metropolitan Life Insurance and just below PepsiCo and Hewlett-Packard. In Los Angeles County alone, the public schools spend nearly $7 billion a year and employ 133,000 people, making it a bigger enterprise than Microsoft, Coca-Cola, or Levi-Strauss (Picus, 1997).

The problem is that in most poor city and suburban neighborhoods, relatively little of the school's funds find their way into the immediate community. The multiplier effect of school expenses on the local micro-economy is severely attenuated and so too are the social capital-generating activities associated with incomes and prosperity.

However, viewing urban communities as assets rather than accumulated deficits provides the means to look at multiplying flows of school funds through such practices as:

- Purchasing goods and services from local producers and suppliers.
- Hiring local residents.
- Targeting contracts for goods and services to support the creation of new businesses.
- Investing resources in local financial institutions, such as credit unions, co-ops, and community development loan funds (McKnight, 1994, 1995; McKnight & Kretzmann, 1993).

Housing

One way to multiply the effects of school expenses is to get school employees to live in communities surrounding the school. Mandates have not been particularly successful in this regard, but housing incentives and career ladder plans that allow para-educators, who now represent nearly one in six school staff members, to attend college and become teachers are one of the potentially powerful incentives for connecting community to schools and connecting school resources to community development (Kerchner, Koppich & Weeres, 1997). Part of the Pomona plan is to develop incentives to attract young teachers to settle in central city neighborhoods. The district has been holding conversations with developers about subsidizing housing for teachers and with financial institutions about making funds available at attractive rates.

Vocational Education

Schools appear most explicitly engine-like when they engage in vocational education. The same industries that seek tax abatements are highly attracted by programs that offer to train their work force. Employer and skill-specific training answers an employer's need for competent entry-level workers and, by using a firm's own equipment for training, targeted education allows training to occur on modern rather than antiquated technology. For educators, the "narrow vocationalism" involved and the apparent movement of benefit from the student to the firm creates an unwarranted public subsidy (Grubb & Stern, 1989). In many cases, school-to-work programs are successful in placing students, but they only boost yearly earnings by $200 to $500 (Grubb, 1996). But if the programs are constructed in ways that allow students to enter the labor market and still have a pathway to higher education, one that has employer encouragement, then an incentive system is created to increase the academic performance of the broad middle range of high school students. These programs work particularly well in cities and use their locational and cultural advantages (Grubb, 1995).

Our industry maps clearly show Pomona as a center of medical practice. A major regional hospital and medical center is located in the city along with many surrounding medical offices and clinics. In addition Western University of Health Sciences is located in downtown Pomona in structures that formerly housed the central retail district. Developing a medical magnet or specialty school becomes an attractive possibility, one that is currently under development.

The district is also planning a business incubator as part of the old shopping center site that is becoming The Village @ Indian Hill. The idea is to support technology transfer and small business development through technical assistance to small and medium-sized firms regarding computer technologies, hardware, budgeting, management skills, and other areas of business.

Education as a Basic Growth Industry

As important medicine is to the city and region, it is dwarfed by the size of education itself. We believe that education in the Pomona area can take on the characteristics of a basic industry, one that spawns other activity. Traditionally agriculture, mining or large manufacturing concerns, such as steel making were thought of as basic industries. In the case of Southern California, and the Pomona area specifically, we suggest that education itself can be an undertaking that creates economic growth and social opportunity.

Basic Industries are Large.
The size of the education industry is important when its employment is compared with that of other Los Angeles County employers. Using data from the California Employment Development Department and the U.S. Census we estimate that employment in the education sectors included in Table 1 makes education the county's third largest industry, behind business management and tourism but substantially ahead of wholesale trade, aerospace, financial services, apparel, and motion picture and television production.[2]

The scale and concentration of education is no less impressive if viewed locally. Map 2 shows the colleges and universities located within ten miles of the Village site and the concentrations of educators living in the region. Notice the concentrations in dark gray, in which there are more than 100 educators per census block group, and the extreme concentrations of more than 500 education workers, in solid black.

There are more than 80,000 college students being taught within 10 miles of the Village site. The Claremont Colleges, which are consistently ranked among

Table 1. Comparison of Education Employment with Other Industrial Sectors in the Los Angeles Metropolitan Area, 1993.

Business/Management Services	419,500
Tourism	264,500
*Education industry**	*234,649*
Wholesale trade (excluding apparel)	229,200
Aerospace	169,600
Los Angeles County public school districts	*133,522*
Financial services	125,800
Apparel/textiles	127,100
Motion pictures/television production	125,300

* Preliminary estimate imputed from employment patterns in public schools.

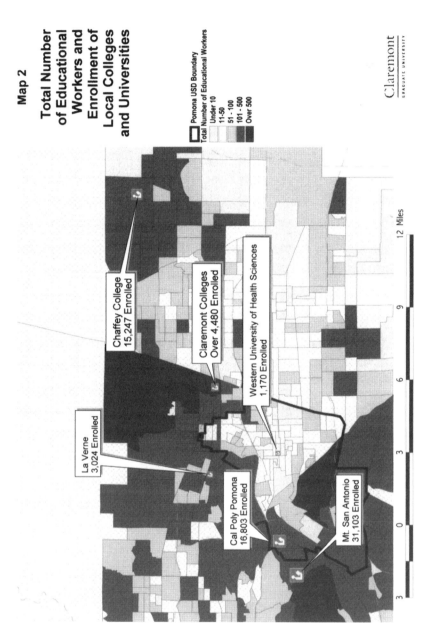

Map 2

Total Number of Educational Workers and Enrollment of Local Colleges and Universities

Pomona USD Boundary
Total Number of Educational Workers
Under 10
11-50
51 - 100
101 - 500
Over 500

Chaffey College
15,247 Enrolled

Claremont Colleges
Over 4,480 Enrolled

Western University of Health Sciences
1,170 Enrolled

La Verne
3,024 Enrolled

Cal Poly Pomona
16,803 Enrolled

Mt. San Antonio
31,103 Enrolled

12 Miles

Claremont
GRADUATE UNIVERSITY

Map. 2. Total Number of Educational Workers and Enrollment of Local Colleges and Universities.

the nation's best private collages, are joined by California Polytechnic University, Pomona, a school considered to be among the jewels of the California State University system. These are joined by two large community colleges with distinguished histories serving the region, Chaffey and Mt. San Antonio, and LaVerne University with nearly a century's service to the region. Together these colleges employee approximately 6,000 faculty, almost all with advanced degrees, and an equal number of staff. In the larger 30 mile commuting range, we find a University of California campus at Riverside, two more CSU campuses, and several private colleges.

As Table 2 shows, over 35,000 education workers live within a 10-mile radius of the Village site. We examined the numbers of education workers, the total population and the student-age population by reference to 1990 census figures, so they understate the current population somewhat, but even using these relatively old numbers, one is able to see the concentration of both educators and students in the area. Taken together these students would form a district larger than many central cities: Miami, Philadelphia, and San Diego among them.

Basic Industries Give Cities or Regions Their Identity and Character.
In some regions of the world, education has this acknowledged characteristic. Universities give Boston is reputation as a seat of learning and attract technological and financial enterprises to the entire region. Proximity to University of California at Berkeley and Stanford University fuels growth in the Silicon Valley. Universities were largely responsible for the startup and growth of the aerospace industry in Southern California. Despite the concentration of educational activity around and in Pomona, the city itself has not been thought of as a seat of learning. Claremont, which lies at Pomona's northern border is pictured as the quintessential college town, but the image never traveled south across the city line.

The Village @ Indian Hill development offers the possibility of creating what the school district calls an educational hub. They are well on their way to attracting multiple educational uses and making connections to colleges and universities throughout the region.

Table 2. Population, Students and Educators Near the Village Site.

	Total population	Student Age Population	Education Workers
Within 3 Miles	211,980	46,432	8,392
Within 5 Miles	416,818	100,990	18,846
Within 10 Miles	917,665	203,494	36,649

Basic Industries Breed Economic Activity.
They send ripples throughout the economy, generating related enterprises and jobs. They also help to create an innovative milieu that replaces obsolete and unproductive industries with new ones. They create synergy.

Part of Pomona's strategy is to create a sufficient critical mass in educational activities that private firms and other educational enterprises will see the area as attractive, much the way that film-related industries congregate on Los Angeles' west side or computer related industries do in the Silicon Valley. By so doing they hope to have a direct effect on the family economies of those living in neighborhoods surrounding the old shopping center site.

SCHOOL AS TRAINING TRACKS FOR DEMOCRACY

In his polemic about American democracy, William Grieder (1992) attests that Americans need 'civil faith' to overcome vast differences among its people. "If these connections between the governed and the government are destroyed, if citizens can no longer believe in the mutuality of the American experience, the country may descend into a new kind of social chaos and political unraveling, unlike anything we have experienced before" (p. 15).

For the past quarter-century, political theory and practice has struggled with the problem of representation and access to school governance. However, interest group democracy generally has not had the palliative effects of increasing the legitimacy for institution of public education that supporters of direct participation in governance had forecast.

Social capital formation begs for a retreat from interest-based politics and toward what Mansbridge calls unitary politics, a politics of shared growth and development (Mansbridge, 1983; Mansbridge, 1990). The basic sense of belonging to something larger than the family and smaller than the nation-state, a sense of mutual dependency, flows from the direct face-to-face politics available only in proximate communities (Bellah, Madsen, Sullivan, Swidler & Tipton, 1991; Etzioni, 1993; Handy, 1994).

In Pomona, as in any urban community, the problem of neighborhood involvement is by no means solved. One beginning effort, which flows from our work with the district, is to raise the perception of Pomona neighborhoods as possessing assets rather as well as liabilities. Poor communities are traditionally seen in terms of their deficits: violence, crime, disease, and dog bites. Community developers and activists have come to understand the limits to a disease-based analysis of neighborhoods, and new tools such as the capacity inventory developed at Northwestern University can direct educators' attention to building on assets, rather than counting deficits (McKnight & Kretzmann, 1993; McKnight, 1994). Recent developments in community indicators, a fusion of environmental, civil rights, and economic concepts promote similar interests (Campbell, 1996; Sawicki & Flynn, 1996). Hence, our four-economy model.

Our focus groups asked residents to mark community assets and liabilities on maps (see Map 3), and the subsequent discussion centered on *their* ideas of the area. (The district is using some of these same elements as it develops educational indicators for its schools.)

Counting assets as well as liabilities allows the school district and others in the community to begin to see their surroundings as something to be built around, not as only a set of problems and barriers. The conversations about assets in focus groups also allowed neighbors to share their perceptions and values with one another. We were surprised with the extent to which respondents, regardless of language, length of residence, ethnicity, or economic status, identified with a need for what the economic development literature calls the "nice" factor. They wanted safe and attractive streets, good commercial areas, parks people can and do use, places people can congregate after dark.

Our interviews clearly showed that the instinct for "nice" is not at all the exclusive property of upper income professionals. The same desire was shared by the people we talked to, many of whom were climbing out of poverty or grabbing a middle-class handhold. Virtually every person who spoke with us asked for safety and decency in residential neighborhoods and commercial streets that were attractive to the eye.

Often "nice" was associated with neighbors and neighborliness. "We're in a cluster of good people, you know, we're like the nucleus of it, you know. And all around it there is all this negative stuff." In situations such as these, even the loss of one person or family is felt, as our participant reported, "There's an older gentlemen here that lives to my right and I told him I was possibly moving to Perris and he said, 'Oh no you don't.' I said, 'Why?' and he said, 'Not until I'm dead and buried . . . I want to keep you as a neighbor.' "

A Lincoln Park resident commented on the virtues of a multicultural neighborhood, saying "It's wonderful. I mean we have such a diversity of people out there . . . We all talk. Everyone is out walking in that area, and it's very positive. You have gay people, Hispanic people, blacks."

A refinement of this same technique would allow communities to identify educational assets in their neighborhoods, and connecting community assets to schooling brings our discussion full circle to the problems of leadership outside the hierarchical triangle.

Map 3.   Assets and Liability Areas of Pomona California.

CONCLUSION: THE LEADERSHIP CHALLENGE

The Pomona school's effort to link its own development and construction with city social and economic development remains a work-in-progress. It's a complex undertaking. As with any urban district, Pomona both thrives on and survives through scores of project initiatives: class size reduction, the Los Angeles Metropolitan Annenberg Project, California's new assessment program, and changes in bilingual education, to name only a few. But even as activity progresses on many fronts, one can easily observe the challenges to conventional beliefs about school leadership.

First, attempting to manage school districts outside the hierarchy challenges existing governance and control mechanisms. To manage the real estate aspects of the Village @ Indian Hill, the school district had to create an independent non-profit corporation. The necessary collaboration between private, non-profit, and public participants made it necessary to lodge governance of the project outside the existing school board.

Second, the job of external relations extends much farther down the school district hierarchy. In a traditional school district, relationships with businesses, city agencies, and other official collaborators is tightly centered on the office and person of the superintendent. In Pomona, as the number of these relationships has proliferated, it has become increasingly difficult to control these contacts and in essence, to control who speaks for the district. The presence of many people perceived of as representing the district has increased the level of uncertainty and at times created tension for the superintendent. As a consequence, a cycle of delegation and recapture of authority has played out over several issues.

Third, the nature of authority delegation itself has changed. In a conventional hierarchy, one delegates task or function, giving a subordinate free range in a circumscribed area of decision making. For this kind of delegation to work, it is necessary that a subordinate understand what is to be done, have the competence to carry out the task, and above all to understand the limits to their delegated powers.

In out-of-the-triangle leadership, idea and project generation takes place at the organization's periphery rather than its center. This means that entrepreneurial activity is often delegated and with it the Parsonian system functions of adaptation and maintenance. In order for these functions to be done without violence to the whole organization, a subordinate must not only be competent, he or she must have a *cultural* understanding of where the district is heading.

Fifth, communications requirements become much more complex. Engaging external audiences – parents, community groups, city officials – is far more difficult than communicating within the school district hierarchy. One of the purposes for developing the GIS maps and data analysis system was to create displays that would be easily understood by both educators and community members. The Pomona experiment is not far enough along to assess the efficaciousness of GIS. Our hope is that graphic displays of complex information will help with two communications difficulties. The first is message omission or distortion in which people don't get information they need. Being 'out of the loop' has been a constant danger. The second danger – too much information – results from attempting to solve the first. Participants report receiving much more information than they can possibly process.

Finally, leadership has simply become more ambiguous. The more the boundaries of the organization are breached and new initiatives are a function of groups and organizations outside, the less leaders can draw on their authority of office to get things done. Leadership qualities of negotiation, idea generation, and a gestalt sense of appropriateness rise in importance.

NOTES

1. Part of the municipality of Diamond Bar is served by the Pomona Unified School District.
2. Other estimates would rank education as the leading industry. Picus (1997) used different industrial aggregations. In his report, finance and real estate is the largest sector with 243,000 employees, making education at least a close second.

ACKNOWLEDGMENTS

The authors thank Superintendent Patrick Leier and former Associate Superintendent Jerry Livesey for their cooperation and assistance in preparing this chapter. Opinions and interpretations of data are solely the authors, and they are responsible for any remaining errors.

REFERENCES

Becker, G. S. (1962). Investment in Human Capital: A Theoretical Analysis. *Journal of Political Economy*, *70*(5, part 2), 9–49.
Bellah, R. N., Madsen, R., Sullivan, W. M., Swidler, A., & Tipton, S. M. (1991). *The Good Society*. New York: Alfred A. Knopf.
Bourdieu, P. (1980). Le Capital Social. Notes provisaires. *Actes de la Recherche en Sciences Sociales*, *3*, 2–3.
Campbell, S. (1996). Green Cities, Growing Cities, Just Cities? *Journal of the American Planning Association*, *62*(3), 296–312.

Clegg, S. R. (1990). *Modern Organizations: Organizational Studies in the Postmodern World*. London: Sage.

Coleman, J. S. (1990). *Foundations of Social Theory*. Cambridge, MA: Harvard University Press.

Cubberly, E. P. (1916). *Public School Administration: A Statement of the Fundamental Principles Underlying the Organization and Administration of Public Education*. Boston: Houghton Mifflin.

Driscoll, M. E., & Kerchner, C. T. (1999). The Implications of Social Capital for Schools, Communities and Cities: Education as if a Sense of Place Mattered. In: J. Murphy & K. S. Louis (Eds), *Handbook of Education Administration*. San Francisco:Jossey-Bass.

Ehrenreich, B. (1990). *Fear of Falling: The Inner Life of the Middle Class*. New York: Harper Perennial.

Etzioni, A. (1993). *The Spirit of Community: Rights, Responsibilities, and the Communitarian Agenda*. New York: Crown.

Flap, H. D., & De Graaf, N. D. (1986). Social Capital and Attained Occupational Status. *The Netherlands' Journal of Sociology, 22*, 145–161.

Fulton, W., Horan, T., & Serrano, K. (1997). *Putting it all Together: Using the ISTEA Framework to Synthesize Transportation and Broaden Community Goals*. Claremont Graduate University Research Institute.

Gamoran, A. (1996a). Do Magnet Schools Boost Achievement? *Educational Leadership, 54*(2), 42–46.

Gamoran, A. (1996b). Student Achievement in Public Magnet, Public Comprehensive, and Private City High Schools. *Educational Evaluation & Policy Analysis, 18*(1), 1–18.

Granovetter, M. (1985). Economic Action, Social Structure, and Embeddedness. *American Journal of Sociology, 83*, 1420–1443.

Greider, W. (1992). *Who Will Tell The People: The Betrayal of American Democracy*. New York: Simon & Schuster.

Grubb, W. N. (1995). Reconstructing Urban Schools with Work-Centered Education. *Education & Urban Society, 27*(3), 244–259.

Grubb, W. N. (1996). *Learning to Work: The Case for Reintegrating Job Training and Education*. New York: Russell Sage Foundation.

Grubb, W. N., & Stern, D. (1989). *Separating the Wheat from the Chaff: The Role of Vocational Education in Economic Development*: National Center for Research in Vocational Education, University of California at Berkeley.

Handy, C. (1994). *The Age of Paradox*. Boston: Harvard Business School Press.

Hirschman, A. O. (1970). *Exit, Voice, and Loyalty*. Cambridge, MA: Harvard University Press.

Hirschman, A. O. (1958). *The Strategy of Economic Development*. New Haven, CN: Yale University Press.

Kerchner, C. T., Koppich, J. E., & Weeres, J. G. (1997). *United Mind Workers: Unions and Teaching in the Knowledge Society*. San Francisco: Jossey-Bass.

Loury, G. (1977). A Dynamic Theory of Racial Income Differences. In: P. A. Wallace & A. LeMund, Lexington (Ed.), Chapter 8 of *Women, Minorities, and Employment Discrimination*. Mass: Lexington Books.

Loury, G. (1987). Why Should We Care About Group Inequality? *Social Philosophy and Policy, 5*, 249–271.

Mansbridge, J. J. (1983). *Beyond Adversary Democracy*. Chicago: University of Chicago Press.

Mansbridge, J. J. (1990). *Beyond Self-Interest*. Chicago: University of Chicago Press.

McKnight, J. (1995). *The Careless Society: Community and Its Counterfeits*. New York: Basic Books.

McKnight, J. L. (1994). Hospitals and Communities Create 'Wise' Environments. *Trustee, 47*(2), 22–23.

McKnight, J. L., & Kretzmann, J. (1993). *Building Communities From the Inside Out*. Evanston, IL: Northwestern University.

Murphy, M. (1990). *Blackboard Unions: The AFT and the NEA, 1900–1980*. Ithaca, NY: Cornell University Press.

Picus, L. O. (1997). The Economic Impact of Public K–12 Education in the Los Angeles Region: A Preliminary *Analysis. Education and Urban Society, 29*(4), 442–452.

Putnam, R. D. (1993). *Making Democracy Work: Civic Traditions in Modern Italy*. Princeton, NJ: Princeton University Press.

Rubinson, R., & Browne, I. (1994). Education and the Economy. In: N. J. Smelser & R. Swedberg (Eds), *Handbook of Economic Sociology* (pp. 581–599). Princeton, NJ: Princeton University Press and Russell Sage Foundation.

Sawicki, D. S., & Flynn, P. (1996). Neighborhood Indicators: A Review of the Literature and an Assessment of Conceptual and Methodological Issues. *Journal of the American Planning Association, 62*(2), 165–181.

Schultz, T. (1981). *Investing in People*. Berkeley: University of California Press.

Tyack, D., & Hansot, E. (1982). *Managers of Virtue: Public School Leadership in America, 1820–1980*. New York: Basic Books.

COMMUNITY BASED ORGANIZATIONS, TITLE 1 SCHOOLS, AND YOUTH OPPORTUNITY: CHALLENGES AND CONTRADICTIONS

Robert A. Peña, Cristal McGill and Robert T. Stout

INTRODUCTION

In order to address a range of educational, social, and institutional crises, states and local communities have looked toward forming linkages across schools and social service agencies to provide needy students and families with coordinated support. Smrekar (1994, 1993) suggests the rationale for transforming schools into houses that educate and provide social service support is founded upon the unique relationship that schools share with students and families. Because schools provide sustained and ongoing contact with students and families, in other words, Smrekar (1994, 1993) suggests they are judged by educators, policy makers, and community members as especially well situated for problem identification and treatment, and for addressing the multiple needs that students and families experience in and outside of schools.

In addition, Smrekar (1994, 1993) and Gardner (1993) suggest that a distillation of discussions and research on school-linked services reveals that policy debates and planning strategies tend to focus on three areas. These areas most often describe the politics of interagency linkages, competition for power

Community Development and School Reform, Volume 5, pages 65–99.
2001 by Elsevier Science Ltd.
ISBN: 0-7623-0779-X

and autonomy, and struggles over turf. Moreover, Smrekar and Gardner forewarn that agencies designed to connect students, families and schools bring a set of assumptions with them about the roles of students, parents, administrators and teachers under this model of education and social service delivery. Finally, these analysts report that despite the changes in roles and relationships that school-linked social service arrangements imply, agencies that actually connect students, families and schools have seldom been researched.

In this study, data describing the community, Raven schools, and Raven School District's collaboration with Friendship House, a community based organization, are provided. In addition, data relevant to the after school and summer school learning programs, and the policies, practices and procedures used in providing for the educational needs of economically disadvantaged children and families in the Raven School District's Family Support Centers (FSC) are introduced. In particular, perceptions of knowledgeable Raven School District and Friendship House students, parents, personnel and staff regarding the formation, purposes and daily operations of social service and education programs and opportunities made available through the FSCs are shared.

These data are provided to fill the void within the school-linked social service literature. They introduce the pressures that helped shape the nature and operation of specific school-linked integrated services while seeking to understand what these pressures implied for operating interagency collaborations and Title 1 schools like those in the Raven Elementary School District. These data and a description of the context surrounding the FSCs are also introduced as scholars assert that understanding how the sociopolitical context interacts with every facet of schools is essential for understanding how successful reform measures may be developed (Bamgbose, 1989; Christian, 1988; Cooper, 1989; Fishman, 1973; Neustupny, 1983; Paulston, 1984; Weinstein, 1986).

THE COMMUNITY AND SCHOOL SETTINGS

The Raven Elementary Public School District (a pseudonym) is located in a county that has grown more than any other in the United States in both the past year and in the seven years since the last census was released. According to state census figures released on March 17, 1998, Maryvale County (a pseudonym) gained 82,789 people from July 1, 1996 to July 1, 1997. In the seven years since the 1990 census, the population of Maryvale County has grown more than 27%, from 2.1 million to 2.7 million inhabitants.

These newcomers to Maryvale are generally of working age and have moved from the Midwest and other states. In addition, the birthrate in Maryvale County is more than twice its death rate and 6,256 immigrants arrived during 1997.

These developments are significant because of their enormity and accelerated pace. Moreover, increases in domestic and international immigration and a greater number of births than deaths constitute the three major contributors to population growth. Reviewing these population trends is important for becoming familiar with some of the numerous pressures that affect Friendship House and Raven school collaborative activities, and the capacity for providers to deliver social welfare and instructional support to students and families in Raven schools.

Located in a low income community in the inner city in the Southwest region of Maryvale County, the Raven School District covers a 6.12 square mile area with 7,856 students enrolled in preschool through grade 8. Since the 1989–1990 school year, enrollment in this district has increased by 42% (see Table 1). The numbers of Limited English Proficient (LEP) students has risen from 1,137 in 1990 to 3,124 students in 1996. These figures represent an 175% increase in LEP students making Raven a majority minority district in terms of the primary language, racial and ethnic characteristics of its students. In addition, the numbers of students receiving free and reduced lunches has grown from 3,168 to 6,136 representing a 94% increase from 1990 to 1996. These trends and recent Title 1 reforms contributed to Raven's being reclassified as a Title 1 public school district while influencing decisions rendered about the goods and services to be provided at the FSCs.

According to Section 1114 of the *Improving America School* Act of 1994, a Title 1 public school district describes an educational agency that is eligible to use Title 1 and other federal, state, and local funds to upgrade its entire educational program on a school-wide level given that for the school year 1995–1996, the

Table 1. Raven School District Student Demographic Information and Trends.

Student Data	1989 –1990	1990 –1991	1991 –1992	1992 –1993	1993 –1994	1994 –1995	1995 –1996	Percent 1989–96
Total Enrollment	5,540	5,789	6,157	6,611	7,128	7,345	7,856	+42%
LEP Students	1,137	1,172	1,518	1,932	2,421	2,638	3,124	+175%
Students Free/ Reduced Lunch	3,168	3,194	3,409	4,693	5,274	5,811	6,136	+94%
Hispanic	37%	44%	52%	59%	71%	76%	81%	+45%
White	51%	44%	38%	31%	19%	15%	10%	−41%
African American	9%	9%	7%	7%	7%	6%	6%	−3%
Native American	2%	2%	2%	2%	2%	2%	2%	Same
Asian American	1%	1%	1%	1%	1%	1%	1%	Same

schools serve an eligible school attendance area in which not less than 60% of the children are from low income families, or that not less than 60% of the students enrolled in the school are from such families. For the school year 1996–1997 and for subsequent years, those criteria changed so that educational agencies serving not less than 50% of children from low-income families, or having not less than 50% of their students enrolled in the schools from such families were eligible to move to school-wide programs and to be classified a Title 1 district.

Section 1114 of the *Improving America School* Act of 1994 allows for public schools and school districts to move from targeted to school-wide assistance so as to relieve them of the need to identify and brand particular students as being eligible for supplemental services and funds. In addition, Section 1114 encourages the developing of more efficient and cost effective schools programs by encouraging education practitioners to secure services from other providers of support including comprehensive technical assistance centers, regional laboratories, institutions of higher education, educational service agencies, and other local consortia.

Altogether, over 700 administrators, teachers and staff are employed in Raven schools. Raven is made up of one preschool, one K-3 school, five K-6 schools, one junior high school for grades 7–8, and one alternative school for grades 7–8. The junior high principal also oversees the alternative school which is housed at the junior high site. The preschool is led by a former principal who is classified as the preschool director. Friendship House's FSCs are located in the five K-6 schools and the one junior high school for students enrolled in grades 7–8.

FRIENDSHIP HOUSE FAMILY SUPPORT CENTERS IN RAVEN SCHOOLS

The first of six Friendship House Family Support Centers (FSCs) located in Raven schools opened in 1987 when the junior high school principal, teachers, staff, and members of the Joseph I. Flores Academia del Pueblo facility organized a food bank and thrift shop in a cluttered stockroom adjacent to the school's main office that was made available to Raven families living in the surrounding community. This school store stocked food, school and household items, and clothing donated by local businesses and community members.

Due to their desire to increase student achievement, and to generate other positive outcomes including increasing the involvement of parents and guardians, a system was developed whereby Raven junior high school students earning high grades for their attendance, academic and behavioral performance were awarded tokens by their teachers and taught to compile a shopping list with their parents at home to purchase items from the school store. Junior high students and families were also provided an inventory of goods that were available for purchase early during the school year, and parents were encouraged to visit the store and school with their children and families.

Three years later, analyses of community newspaper reports and school district archival data revealed that pressures exerted by Raven's school board, administrators and teachers, the local community, and a change in the superintendency coincided with the discarding of a token system that seemed too difficult to standardize, and to the school store being changed. In addition to making donated goods and clothing available at no cost to Raven families, the school store was renamed a "family support center" or "FSC." Individuals working in these centers were responsible for welcoming new parents and families to Raven schools while providing health screening, information about adult education programs, social service and prevention activities, youth programs, health care services, home and personal care, rehabilitation, preschool, and parenting skills training.

Later, as the FSCs spread to five other Raven schools, parents as partners programs were created, comprehensive social and education programs to address the problems of substance abuse and gang activity were added, and Friendship House's after school and summer school drop out prevention programs were developed. The changes mentioned above also resulted in Friendship House personnel and staff becoming more visible in Raven Schools, and in two individuals hired and paid by Friendship House being located in each FSC to coordinate and oversee the services and goods that were provided there.

While Raven schools' six FSCs often resembled a lounge with rugs, couches, over stuffed chairs, bookcases, hanging artwork, and a few study carols located to the front of the room, they also looked liked mom and pop stores with cola and orange soda, bags of potato chips, cereal, cookies, laundry detergent, brooms, dustpans, mops, and other goods organized in rows on metal wire shelves behind a long wooden counter with a laminated countertop. Most often, FSCs also resembled thrift stores with shiny metal closets for hanging sweaters, blouses, skirts, slacks, shirts and work clothes positioned toward the left, and boxes brimming with shoes and young children's clothes that were clean, pressed, well cared for, neatly folded, and occasionally in a heap on tables positioned toward the back of the room on the right.

FRIENDSHIP HOUSE AND FAMILY SUPPORT CENTERS: BACKGROUND INFORMATION

Friendship House was created in 1920 as part of a local initiative by the Phoenix Americanization Committee and the U.S. Department of Education to provide social welfare support and to promote literacy among new families entering the United States. During the 1980s, Friendship House enhanced its visibility, influence, and comprehensive services by purchasing a building and creating the Joseph I. Flores Academia del Pueblo. This facility and its programs were originally designed to provide positive alternative activities for inner-city youth. In addition, these programs were intended to promote educational, social, and cultural enrichment activities in collaboration with Raven and other public school districts located in South Phoenix Arizona by sharing facilities and establishing school-based family support centers.

According to a mission statement drafted in 1928 and most recently revised in 1992, Friendship House was created to serve children and families by helping them address their social welfare concerns, and by helping elementary and middle school aged youth improve their academic performance in school. To that end, Friendship House developers anticipated the need to both provide families with adult education programs, social services and prevention activities, youth programs, home and personal care, rehabilitation, preschool materials, parenting skills training, and parents as partners programs, and to provide academic and other types of social welfare support by collaborating with Raven and other neighboring elementary schools. In addition, in 1976, the Phoenix South Community Mental Health Center and the Phoenix Revitalization Corporation joined the Raven-Friendship House interagency collaboration and influenced the developing of the After School and Summer School learning programs in Raven and other school sites.

Friendship House Provides Instructional Support

Friendship House's After School Drop Out Prevention program was developed in 1979 to help students realize higher academic outcomes in school. This program met Monday through Friday after school from 3:00 to 6:00 p.m. Trained Friendship House instructors and tutors provided students with a snack, assistance on daily homework assignments, instructional support, and transportation from Raven school FSCs to the students' homes. Instructional support and assistance with homework assignments were intended to be provided to students in grades 5 through 8, although siblings and Raven students from earlier grades were routinely present during after school instruction.

Friendship House's Summer School Drop Out Prevention program was also formed in 1995, but developed, instead, to encourage and enhance students' motivation to learn. This program met Monday through Thursday during June and July from 8:30 a.m. to 2:30 p.m. in the FSCs and unused classrooms in Raven schools. Certified Raven teachers, student-teachers and tutors trained at the local state college were hired and paid using Friendship funds to introduce and instruct students using Raven School District and Caesar Chavez Academy curricula. This summer school instruction was to focus on enhancing the students' skills and knowledge in reading, writing, math and science. The Caesar Chavez Academy curriculum was developed in cooperation with local state university faculty and, according to Raven and Friendship House staff, with the pupils' cultures and background interests in mind.

Lunch and transportation from their homes to the summer school programs offered in Raven schools were made available and provided to students using local donations, state, and federal funds obtained by Friendship House's director. Additionally, donations and funding afforded field trips to local Arizona events and landmarks for Raven students in grades K through 8 who signed up for summer learning.

THEORETICAL FRAMEWORK

The framework utilized to study the relationship between Raven schools, students and families and the FSCs was also used to develop the protocols and to guide data collection. It was developed by combining urban school restructuring and student dropout prevention literatures. In their research on urban school restructuring and student dropout prevention, Wehlage et al. (1992, 1989) proposed to develop new knowledge on how organizational features of urban schools could be changed to improve education for different students including students from economically disadvantaged backgrounds.

Wehlage et al.'s (1992, 1989) research was founded upon syntheses of prior knowledge, new analyses of existing data, and new empirical studies of public elementary, middle and high schools. It was utilized to conduct this study because of its focus on organizational, school restructuring, urban, and student at-risk issues. Further, Wehlage et al.'s (1992, 1989) framework was utilized here as it was proven to be effective for exploring the nature of school, family and interagency relationships in previous studies (see Wehlage et al., 1992, and Peña, 1995, 1993).

Wehlage et al.'s (1992, 1989) framework reflects four distinct themes, each with its own programmatic efforts at change. The first addresses the nature of student experiences in and to a more limited extent, outside of schools. It is concerned with the quality of curriculum, instruction, assessment, school climate, discipline, and student support provided in academic and nonacademic areas. The second theme is concerned with the professional life of teachers. It considers the range of new roles and responsibilities that define teachers' work.

The third theme focuses on school governance, management, and leadership. This category refers to ways in which authority and accountability are allocated. It also calls for creating new mechanisms for making decisions that involve sharing power with others in the community.

Wehlage et al's (1992, 1989) fourth theme calls for finding ways for urban schools to draw upon community resources to enhance the chances for disadvantaged youth to achieve school success. It came from analyses of urban school efforts that involved integrating and coordinating health and social services for children and families, programs for youth employment, incentives and mentoring for higher education, as well as from analyses of research that called for infusing private sector resources into the curriculum and other academic activities that students experience in schools. Similar to ideas introduced by Gordon, Curran and Avila (1966), Hobbs (1982, 1979, 1978, 1975, 1966), Gardner (1993), and Smrekar (1994) in their discussions of schools, needy students and families, and interagency linkages, Wehlage et al's (1992; 1989) fourth theme is founded on the assumption that utilizing different resources from surrounding communities can assist in providing youth with the additional support they need to succeed academically in school.

RESEARCH DESIGN AND METHODOLOGY

Research Design

To examine the nature of the relationship between the activities and services provided by the FSC's on the one hand, and the experiences of Raven schools, students and families on the other, I relied upon a research design comprised of hermeneutic, empirical, and critical data collection methods (see Fig. 1). This design was utilized because of its focus on organizational, school restructuring, urban, and student at-risk issues. It was also utilized as it was proven to be effective for exploring the nature of school, family and interagency relationships in prior research (see Peña, 1995, 1998).

Hermeneutic data collection methods seek to discover meanings that individuals attach to features of their environments "to provide a deeper understanding of the context in human terms" (Sirotnik & Oakes, 1986, p. 81). These methods of research are oriented toward the interpretation and understanding of social events "in terms of the participants in those events including the researcher" (p. 23). Researchers suggest that hermeneutic data collection methods allow investigators to discover how "meaning about the organization is intersubjectively communicated and created" (Foster, 1984, p. 255). These methods were selected for this investigation based upon the

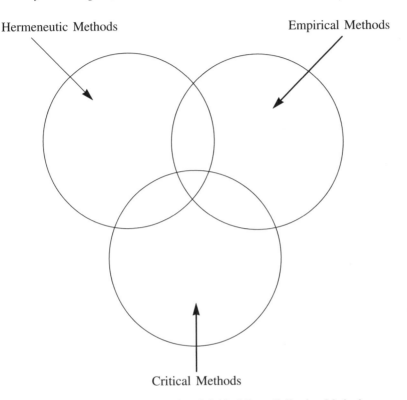

Fig. 1. Hermeneutic, Empirical and Critical Data Collection Methods.

assumption that interagency designs that link schools to other organizations are political and dynamic structures.

Empirical data collection methods find their origins in the natural sciences. These methods place a premium on explanation and involve the "systematic and ongoing collection of descriptive data from participants about features of the school context" (Sirotnik & Oakes, 1986, p. 81). They require generating a comprehensive knowledge base to compare and contrast with features of schools and the connections between Raven and the FSCs. They also involve the collecting and analyzing of monographs, articles, texts and current research on interagency collaborations that connect community based and other organizations with schools. Empirical data collection also calls for developing a conceptual framework that identifies issues pertinent to forming effective school and community collaborations.

Critical data collection methods seek to discover "why things are what they are, how they got that way, and whose interests are being served by existing conditions (Sirotnik & Oakes, 1986, p. 81). These methods of analysis place a premium on mining and analyzing discourse and behavior to obtain "clarification of values and human interests" (p. 19). The purpose for their use is to identify unequal social relations as they occur in institutions. Critical methods of analysis were used during this investigation to address questions about the nature of the linkage between Raven schools and Friendship House's family support centers, and to determine what this linkage implied for Raven's Title 1 schools, and their neediest students and families.

Research Methods

Wehlage et al.'s (1992, 1989) theoretical framework and the hermeneutic, empirical, and critical design shown in Fig. 1 informed the procedures utilized in this study. I began by using qualitative research methods, collecting documentation, and observing meetings where Raven and Friendship House administrators, teachers, tutors, and staff talked. Simultaneously, I learned about the goods and services that were provided, and observed different activities and interactions that occurred in the FSCs. Data derived by using these methods led to my developing questionnaires that included categorical and quantitative items. These items were written using a critical theoretical perspective, and administered to Raven school and Friendship House personnel. Questionnaires were also distributed to Raven students and families.

Hermeneutic Methods

Raven and Friendship House personnel met once every three months in each of the six individual school settings for 30 to 90 minutes during the three year period when data was collected. I watched on 14 occasions as they discussed the operations, successes and failures of the FSCs to obtain resources and involve different personnel, students and families. I also took field notes as these personnel discussed the goods and services they provided, and how to improve their communication and instructional practices. Some Raven and Friendly House personnel also met individually and, on occasion, with the associate superintendent. Some spoke at board meetings. I observed 14 school meetings, 3 of 6 meetings with the associate superintendent, and one school board meeting, and slightly over 127 hours of observation data were logged Next, I interviewed 48 participants to amass a breadth and depth of information on the subject of the relationship between the schools and FSCs. Each of the interviews lasted between 30 minutes to two hours and was recorded and transcribed. One trained research assistant, two principals, one Raven teacher, and Friendship House's director assisted in developing the interview protocols. I also used the theoretical framework to develop and organize the interview questions, and to develop protocols for observation and document analysis purposes.

The questions I posed were open ended, borne from the theoretical framework and designed to elicit the participants' views and understandings about Raven schools, the FSCs, their activities and use by Raven students and families. I started by asking each individual to recall what they "experienced," "understood," and "felt" about the support centers. We discussed how they operated, and the services the centers provided to Raven students and families. I asked these participants to tell, "as best they could," how often they were involved, how they participated, and how they were supported and limited in their efforts to participate. In addition, the participants were asked to "assess the quality of the services and goods that were provided," and "to describe characteristics of the schools, district, and Friendship House that hampered their opportunities to benefit and to provide benefits to others."

Prior to and at the start of the interviews, I explained the study was intended "to understand how the schools and the FSCs worked together" and "how their combined activities benefited different personnel, students and families." I assured each of the participants that that their responses would not be associated with their names, peers, or affiliations. I explained that "no harm would come to them" through their involvement in the study and this seemed to encourage them to be forthcoming with their assertions. Most often, I met with the educators on school grounds and Friendship House personnel in the FSCs, and at the Joseph I. Flores Academia del Pueblo facility.

Constant comparison (Bogdan & Biklen, 1992; Glaser & Strauss, 1967; Glesne & Peshkin, 1992) was used to analyze the data I collected and to form categories that captured the perceptions of the participants. Later, 14 participants agreed to read, edit and verify the accuracy of the categories that I developed given their accounts, accounts provided by other participants, and given the data I compiled during observations and document analyses. These 14 participants, including four principals, the Friendship House director, three Raven teachers, two instructors, three students, and two parents were also involved in assessing the accuracy of the categories that were formed given their participation in the study, the influence of the school setting, the different components of the revised theoretical framework, and my presence as the researcher.

Those assertions made by the participants that were repeated often and taken from each of the data sources were judged by the 14 participants and myself

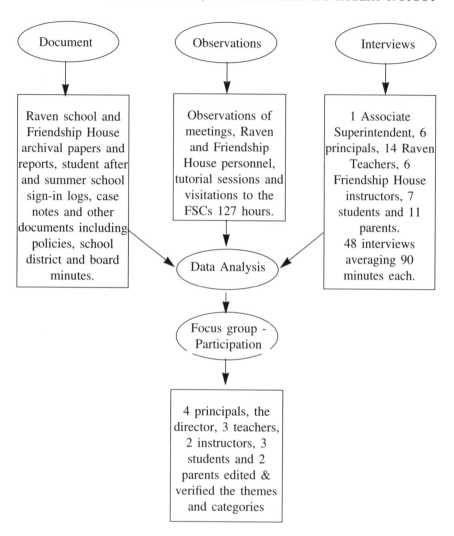

Fig. 2. Qualitative Methods and Procedures.

to be the most reliable for developing and redeveloping categories. These categories were, once again, examined by the participants and myself individually and during two focus group sessions held in a meeting room at district office to account for patterns found across frequent and rare events, to account for similarities and differences across confirming and disconfirming evidence (Bogdan & Biklen, 1992; Glesne & Peshkin, 1992), and to add validity to the conclusions drawn from analyzing the data.

Actual quotes taken from different data sources were used for coding chunks of data. Miles and Huberman (1994) define coding as data analysis, indicating that the meaningful dissecting of data involves keeping the relations between the parts intact. Analyzing the data also called for using critical theoretical perspectives as they were found to be effective for understanding how organizations like schools collaborate with community based organizations and deliver services to students and families in previous research (Capper, 1994, 1997).

Empirical Methods

As with the hermeneutic methods already described, the theoretical framework and the data collected using qualitative procedures informed the developing of the questionnaires. In particular, items included in both English and Spanish on two separate but otherwise identical questionnaires were stapled together, and divided into categorical and quantitative sets. The categorical items invited respondents (in English and Spanish) to identify their affiliation with either the Raven schools or Friendship House, and to classify themselves as an administrator, teacher, staff member, student, or parent. Respondents were also asked to identify their native language (English or Spanish, or Other), and the language they used most at home (English or Spanish, or Other). Additionally, respondents were asked to indicate "yes" or "no" describing whether they did or did not participate in Raven school's free and reduced lunch program; whether they visited the FSCs on from "1 to 5," "6 to 10," and "10 or more" occasions, and what goods and services included on the questionnaire and available at the FSCs that they experienced and were involved in providing to others.

Quantitative items invited the respondents to rate the quality of the goods and services they obtained and/or played a role in providing through the FSCs. Respondents were asked to indicate how satisfied they were, with a 5 indicating a very high level of satisfaction, 4 satisfaction, 3 average, 2 no satisfaction, and 1 very bad. Raven and Friendship House administrators, teachers, instructors, tutors and staff were asked to identify and rate the quality of only those goods and services with which they were involved in providing to other personnel, students, and family members. Students and parents were asked to rate only the goods and services they obtained as being of high, much, some, little, and very little satisfaction. In addition, all respondents were asked to place a zero (0) next to those item(s) that they were not involved in providing to others, and that they did not obtain or experience.

In sum, 600 5th through 8th grade students completed questionnaires, although only 138 indicated that they visited the FSCs on "6 to 10" and "10 or more" occasions. Of this number, 112 questionnaires were appropriately completed and analyzed for this discussion. In terms of families, 97 parents and guardians representing an equal number of different families returned completed questionnaires, and of this number, 86 were analyzed. Over 300 middle level Raven school educators returned completed questionnaires, and of this number, 169 identified interacting with the FSCs on "6 to 10" and "10 or more" occasions and were analyzed. All 21 Friendship House personnel who were interviewed, also completed questionnaires that were analyzed for this study.

The Raven school educators whose questionnaires were utilized included 17 administrators and 169 teachers. Friendship House personnel completing questionnaires included the director and all 12 coordinators who were classified as administrators and responsible for overseeing the goods, services, and coordinating the activities provided by the FSCs. Four Friendship House instructors and four staff/tutors who aided in distributing goods, services, information and were responsible for providing after and summer school instruction also completed surveys.

FINDINGS AND DISCUSSION

To understand the logic and to establish the validity of the categories that emerged from the critical analyses and participant checks of the data collected using qualitative methods, excerpts are included. These excerpts are provided as evidence of findings gathered during the data collection process and are organized in relation to the different components of the theoretical framework. In addition, excerpts are provided to maintain the integrity and voice of the participants and to enable readers to form their own interpretations (Capper, 1994).

Excerpts taken from the data compiled using qualitative procedures are also given to confirm the importance of the participants' experiences and views on how the Raven schools and the FSCs worked together, and to gain their perspectives regarding how their combined activities benefited different personnel, students and families. A fuller description of this study was provided in the form of a report to Raven School District personnel. Upon its completion and with the participants' consent, a copy of this report was submitted and is available in the form of a monograph from the Resources in Education Clearinghouse, No. EA 424 649.

Data compiled using quantitative techniques provide descriptive information about different respondents, and their relationship(s) to the FSCs. These data also describe usage patterns, and levels of satisfaction that individuals had with the different goods, services and instructional practices that the FSCs provided. This information is presented in different forms, including frequency tables and a pie chart where appropriate. It is integrated and reported in tandem with data collected using qualitative approaches, and organized according to the different components of the theoretical framework. This presentation of findings and discussion is followed by a discussion of how organizational and other features of urban schools might be changed to improve education for students from economically disadvantaged backgrounds, and Title 1 school and community relations.

Participant Membership: Respectful but Uneasy

According to Wehlage et al. (1992, 1989), for membership to occur, personnel need to develop positive and respectful relations with each other, with students and families. In this study and in other research on school restructuring and student dropout prevention (please see Peña, 1998, 1995, 1993), participant membership describes efforts made by administrators, faculty and staff to develop relations that are positive and based upon mutual respect, and to create positive and respectful relations with students and families. Developing participant membership also calls for providing goods, services, counseling and instructional strategies that are perceived to be useful and relevant for addressing the current and future needs of different individuals. In addition, Wehlage et al. (1992, 1989) explain that developing participant membership involves individuals in expressing care and concern for others.

"Respectful but uneasy" was the statement offered by one Raven educator that focus grouped members identified as describing the alliance among Raven and Friendship House personnel best. Raven educators respected the access and knowledge that Friendship House personnel acquired through their visits to the homes and communities of the students and families, and they described their relations as positive, but these individuals were also uneasy about many of the services provided in the FSC, and not optimistic regarding the capacity for Friendship House instructors and tutors to deliver adequate instructional support. On the other hand, Friendship House personnel were respectful of the administrators' and teachers' knowledge and training, but they held doubts about the commitment, level of caring, rigidity and willingness of Raven educators and staff to share the information, resources, and time that was essential for supporting students and families.

Analyses of responses to questionnaire items indicate that higher percentages of students and family members perceived their relations with others in the

FSCs and the Raven schools favorably. Larger percentages of Raven and Friendship personnel including instructional and administrative staff perceived their relations with one another to be average however (see Table 2 and Table 3 in Appendix A).

In this study, Raven and Friendship House personnel were uncertain and generally divided about crossing the boundaries that separate schooling and family life. Raven administrators, teachers, and staff demonstrated a penchant for focusing on academic matters and the intellectual development of students. For them, appropriate family involvement was defined in relation to parents and guardians trusting and generally not interfering with their decisions about education. It also was founded upon the willingness of parents and guardians to be "helpful," "visible," and "supportive" in classroom and school functions.

This approach toward separating the developing of the brain from other factors that affect the development of students seemed due to how the educators were trained, how they defined their professional responsibilities, to their years of experience in and outside of Raven schools, and to cultural factors and the expectations they held about students and families. Cultural factors and participants' expectations are described later in this study.

Commenting on how they positioned students and families to obtain goods, services and instructional support, one teacher's explanation captured sentiments and behavior patterns expressed by a majority of administrators and her teacher colleagues. This teacher described "feeling good" about the FSCs, and "the willingness of teachers and the district to give students and parents things that help them with their problems." She added that she felt "proud because many other schools don't show [students and families] caring like we do," but that she also felt "guilty" because she "[did not] always tell [students and families] about the tutoring at the FSCs."

This teacher and other educators explained that they preferred to "help the ones who need help with learning by myself," that "I know some of them are in school and being trained, but they're awful young and many [Friendship House instructors and tutors] are not well supervised," "just starting out and don't really understand what we need to do." These teachers added that they were "trained" and "accustomed to doing the teaching part," "unprepared" and "drafted" to involve the FSCs, and that they felt nervous and uncomfortable about being perceived as "dishing off our teaching responsibilities to persons outside of the students' families." In contrast, new teachers and teachers just starting out in the district seemed less knowledgeable but more eager to work with the FSCs, and more convinced that the FSCs were an integral part of Raven schools.

A majority of Raven teachers also expressed dissatisfaction with existing teacher evaluation practices, and especially their levels of success in educating students being judged using policies, and procedures that did not account for the role that Friendship House instructors and tutors played. A few added that Friendship House tutors "did the students' work themselves," and "give too much help," and they introduced concepts and methods for problem solving that the teachers did not want the students to experience. Time and resource constraints were also reasons that Raven educators gave that fueled their resistance to involving Friendship House instructors and tutors in providing academic and instructional support.

Friendship House administrators, instructors and tutors were skeptical about the commitment, level of caring, rigidity, and willingness of Raven educators and staff to share information, resources, and time due to difficulty they experienced in exchanging assignments, information, and meeting with teachers, and because of differences in expectations and job status. One instructor described obtaining a mathematics textbook but only after making numerous attempts, and having the teacher "show me in the front where it says the book is the property of Raven schools and not Friendship House or even the students." Others described the requirement to "sign out books and other stuff" making them feel as though they were not trusted, and the using of their "own money" to purchase "supplies" and "chapter books even though [the teachers] have some they use inside the schools with the kids."

In terms of their perceiving they had different roles and unequal status, Friendship House administrators, instructors and tutors suggested it was obvious to them that Raven personnel interacting with the FSCs preferred discussing "teaching and learning" and "where they are with their planning," and that teachers and administrators generally "disrespected," "did not ask questions" and did not "seem as knowing" and "as interested" in matters related to families and communities. Analyses suggested that having the FSCs provided Raven administrators, teachers, and staff with the opportunity to use Friendship House personnel as family and community brokers. This was evidenced as numerous Friendship House personnel recalled Raven principals and teachers becoming involved "not to ask about what they're learning, but mostly when they want to know what's happening with the parents or inside of the home of one of the kids."

This finding differs slightly from previous research (Smrekar, 1994), suggesting that unless strategies for involving agencies in mediating relations between schools and families are well thought out, deliberately formed, and carefully implemented, community based agencies like Friendship House and Raven's FSCs are likely to foster practices that not only reduce the frequency with which educators and families interact, but that also lead to classifications that place the importance of addressing family social welfare needs, and cultivating positive and respectful relations with students and families on other than academic lines at risk.

When asked to comment on the quality and nature of the goods, services and instructional support provided in the FSCs, and the interactions between Friendship House and Raven personnel, the students' and parents' responses mirrored those given by Friendship House personnel. Students and parents described a greater willingness and tendency to discuss personal problems with FSC instructors and tutors, and the feeling that while Raven teachers cared, they placed greater importance and were mainly concerned with education issues.

Students and parents described FSC personnel as "members of community" and "experts about the neighborhoods," and Raven administrators and teachers as "smarter about learning" and "doing the learning things." One student described the FSCs as providing goods and services that "you sometimes get in the churches where you live," and Friendship instructors as "not as smart as our regular teachers." A parent who was new to the district explained that FSC personnel "speak Spanish" and "help you with getting what you need today and the schools always helping you get what you need for tomorrow."

A third parent described the director of Friendship house and his wife as "our surrogate family because they're with you and show you around when signing up for getting assistance and they give you their [telephone] number, tell you where they live and help with the everyday things." Many students and parents also sensed that Raven teachers were less knowledgeable and interested in the FSCs and their neighborhoods, and that they were less willing to discuss family concerns. This led some to feel "embarrassed" and "made to feel out of place," when family issues intersected with teaching and the learning of students in school.

When asked to describe how Raven and Friendship House personnel interacted, the students and parents not only suggested that Raven administrators and teachers "cared more about doing things with the books and the FSCs cared more about the communities," but they indicated they preferred that the opportunities that they experienced were separated along Raven and Friendship House lines. Students and parents explained this division of services and responsibilities helped them to "know where to go" and "who to talk to for what you need," and that it enabled them to "keep family business and school business private" and "not bring people who don't need to know about what you're doing with the family and the other people who don't need to know school things together."

Participant Engagement: Unclear and Constantly Shifting

Participant engagement is generated through providing a variety of goods and services, and developing multiple counseling and instructional approaches to increase the involvement of Raven and Friendship House personnel, students and parents. In addition, it involves delivering goods, services, counseling, and instruction that are likely to result in group, individual, and personal success (Wehlage et al., 1992; 1989). Participant engagement also requires that personnel are alert and that they work to eliminate policies, procedures, and practices that may separate, disadvantage, and otherwise stigmatize individuals according to class, race, ethnicity, and other lines.

Analyses of responses to questionnaire items indicate that significantly higher percentages of students and family members utilized social welfare services than they did after and summer school instruction in the FSCs (see Figure 3 in Appendix B). Students and families also rated the goods and services provided in the FSCs more favorably than Raven and Friendship House personnel. Specifically, a majority of Raven and Friendship personnel rated the goods and services provided as average and below average (see Table 4 in Appendix C).

According to one principal who had worked in the district "for over twenty-five years," the FSCs had "strengthened and improved" Raven's overall relationship with Friendship House and the surrounding community. "Turnover in personnel," "changes in the local neighborhoods," "minority flight in and white flight out of the community," "changes in the board," and "changes in district office personnel" adversely affected the developing of sustainable engagement, on the other hand, and according to Raven and Friendship House personnel, resulted in the potency of the FSCs, Raven's and Friendship House's linkage "raising and declining from year to year and from school to school." In addition, Friendship House personnel reported rates of donations and volunteerism as varying and having implications for the breadth and quality of services provided which, in turn, "affects what we can do, who, how many, and probably why or even if students and families come to the FSCs."

This constant sense of flux swirling around the schools, Friendship House, and the surrounding community gave rise to confusion, frustration, disappointment and despair. For example, both Raven and Friendship House personnel described dependable relationships with students, family members, and members of the community as being "infrequent" and "very rare." Raven administrators and teachers recalled "meeting people" and "making plans" with individuals who later "could not continue" to participate or "said they couldn't play a long term role." Friendship House personnel described routinely competing with other community based organizations for limited funding and community support, and finding that the capacity for local community members to contribute and become engaged was "all tapped out."

Raven and Friendship House personnel also described addressing "needs" and "problems" as a primary responsibility of the FSCs, suggesting that once treatment was administered and problems were addressed, the need for personnel to maintain contact with each other, students and parents over specific issues would "go away." They identified obtaining information, goods and services that were relevant and effective for engaging school and community support, and securing the participation of students and families as "two of our biggest challenges."

Raven administrators, teachers and staff also described being "not optimistic," "somewhat hopeful," and "not hopeful at all about the centers or anything we do having a longterm effect." "What we have in the centers," one teacher explained, "depends on what people give and what we can get from people who live around here, keeping in mind that most really don't have much to give other than problems and their needs which have to be addressed, and that those who have something to offer usually leave once they get the chance." Other Raven and Friendship House personnel described their remaining involved by "remembering the smaller successes," and "doing what you can to help a few students and families make it."

In contrast to those views provided by Raven and Friendship House personnel, students and especially parents and guardians were appreciative of the FSCs, and Raven and Friendship House "helping with making a difference." Students and parents used terms like "home base," "our other place," and "a safe and clean place," to describe the FSCs. Parents and guardians confided that "you don't worry," and "you know your kid will be doing good stuff," "protected," and "safe in the center."

Students indicated that individuals working in the FSCs emphasized "good citizenship skills," "avoiding crime and drugs," and "students completing high school." One student explained that FSC staff "want us to fit in good and be successful with other good people." Another student commented that "they're [instructors and tutors] always saying about watching out in the neighborhood, and staying away from drugs and people who are trouble all the time." Three students agreed, in separate interviews, that school practitioners and Friendship House staff suggested that the FSCs "was for making sure you finished [high school]," "graduate," and "finished twelfth grade so you could get a good job and get some money and a family and whatever." Each of the seven students interviewed also indicated that practitioners and staff emphasized "cultural awareness."

These students perceived that Raven school practitioners and Friendship House personnel felt that cultural awareness was "important."

Four students explained for example, that "the instructors and sometimes the tutors talk about Mexico," and "how things were like back in Mexico." "They said that in the United States you have more freedom to do stuff and to make more money than in Mexico," one student explained. Three students described "how school teachers said it was good that we could speak two languages, but that we would have to speak English really good to do good in the United States." Finally, two students added that Raven school and Friendship House staff said "it would be hard," and "a lot of people don't think we can make it, but that we should keep on trying, and keep working hard." These students explained that Raven and Friendship House personnel "said we should be proud," and "[not] forget our families and where we came from," suggesting that "them making you feel good and like you belong here makes the students want to come."

Participant Beliefs: Fighting the Fight and Not Letting the Ship Pull Away

Numerous at-risk and organizational reform studies from a range of disciplines emphasize commitment. In this study, participant beliefs is measured according to the willingness and capacity for Raven and Friendship House personnel to sustain their commitment to achieving those objectives of the FSCs related to providing students and families with social welfare and academic support. In addition, participant beliefs describes personnel expressing commitment to working with one another, needy students and families; to their extending their roles to help different individuals with personal problems, and to Raven and Friendship House administrators, faculty and staff believing that they and others can succeed. Finally, participant beliefs describes Raven and Friendship House personnel not giving up when they meet resistance, and to their having tolerance and appreciation for attitudes and behaviors that are different, challenging, and unexpected (Wehlage et al., 1992, 1989).

While critical analyses of the data collected using qualitative techniques suggests that the optimism of Raven and Friendship House personnel was severely tested and that many possessed scant hope for the long term futures of their students and families, the commitment, patience, and professionalism these individuals showed toward working together and carrying out their assigned roles and responsibilities to others was strong and ongoing (see Table 5, Appendix D). This commitment and ethic of caring was provided despite numerous school, community, family, and personal limitations that cuffed their potential to help Raven students and families to succeed. School and FSC personnel often likened their experiences to survival. They compared their activities to a "tug-of-war," suggesting that delivering education and social welfare support was like "not looking down," "holding on by your fingers with some others," and "not letting the rope [on a ship] slip or get away."

Raven and Friendship House personnel also mentioned administrator and faculty turnover, suggesting that "working in poor schools and neighborhoods," and "staying in schools like Ravens is not always possible for people who aren't committed," "don't want to get dirty when they work," or "for persons who have a weak constitution." Taken together, analyses of these and other similar data also suggested that the willingness and efforts made by Raven and Friendship House personnel to participate, not give up, and get involved in providing a range of support services, also frustrated their efforts to assess the quality and nature of the effects of the goods and services they provided. Many teachers described being "unsure" about their students regular participation in after and summer schools learning, and a majority stated that they "did not know," and were "uncertain" regarding whether the FSCs were promoting achievement and having other desired effects.

Similarly, a few Friendship personnel described the need for collecting data and conducting assessments and follow-up activities. Lacking time, knowledge, resources, and skills, these individuals assumed, instead, that the goods and services provided in the FSCs were having the desired effects. These individuals described providing immediate and short term assistance, and "hoping but not really being sure if it improved [the students and members of the families] behavior." They also hoped and perceived that when the visitations made by students and families stopped, and when communication between Friendship House and Raven personnel declined, this stood as evidence of success. One Friendship House instructor stated that "you assume if [the students and families] stop coming for help, they must be doing better." Another reasoned that "we sort of have a philosophy that if it ain't broke, then don't spend money and time trying to fix it." She later explained that

> our best cases happen when the teachers contact us, or when some other group or agency or church member calls us to see what we can give in the way of counseling or maybe food or clothes or other help. When and if the student or brother of a family comes, you do all you can and you don't turn people away. When they stop coming, or the teacher or whoever doesn't follow-up, then you assume everything's better and that you were able to make a difference.

Analyses of the data collected through interviews, interactions and observations of students and families in the schools and FSCs, and the study of students' class work indicated that each categorized the commitment of Raven and Friendship House personnel involved in the FSCs as "very strong" and "strong." Words and phrases most often used by students and family members included "always helpful," "real interested," and "involved from their hearts." Observations of students during class and lunch hours and while they participated in the FSCs similarly indicated that they enjoyed positive, supportive, and what appeared to be ongoing relationships with other students and with Raven and Friendship House staff. Data describing interactions outside of the schools between different students and families were not routinely collected.

No discernible pattern of isolation and alienation was apparent in this study as student and family interactions in the FSCs crossed genders, languages and age groups. In contrast, indications that student differences and cultures were acknowledged, accepted and appreciated were revealed through analyses of student class work including some student journals. In particular, analyses indicated that discussions and descriptions contained in students' work included positive phrases, self-images and reflections of interactions with other students, Raven and Friendship House staff. Raven and Friendship House personnel routinely communicated with different students and families in writing using their native languages, and commenting, for example, about their desire to communicate regularly, to have parents "sign work," "telephone," and "get more involved."

Further, evidence of the commitment shown to students and families emerged when they were able to recall occasions when they were praised, recognized and made to feel welcome by other students and Raven and Friendship House administrators, faculty and staff. One student described how a staff member frequently called her "mia (mine) like my mom does." Another recalled "when [a FSC staff member] asked me to help [another student] with his fractions, then gave me a candy before we left. This made me feel good like he knew I could do it and he really cared." In addition, parents and guardians from five different families explained how FSC staff "just talked to you like you were important," how staff "knew things," "listened," and "didn't always have to say something when you talked to them." These parents and guardians added that Raven and Friendship House staff "even talked with you about things that was going on and didn't have nothing to do with what was going on over here [at Friendly House]." They recalled that staff "even remember what you told them the time before, so you know they must be interested in helping you out."

Raven School and Friendship House Support: You Never Know What You Might Get

Raven and Friendship House support describes the degree to which administrators, faculty, staff, students, parents and guardians have autonomy and power to make decisions that affect the FSCs. It calls for allocating a sufficiency of time and resources for developing the centers, and for enabling different individuals to enhance their skills, knowledge, and interpersonal working relationships (Wehlage et al., 1992, 1989). Raven school and Friendship House support, in this study, also describes the providing of opportunities for personal and professional development that dovetail with increasing the capacity for individuals and the FSCs to provide relevant and ongoing support.

Analyses of many of the findings already introduced suggest that the nature and operation of the schools and centers were governed, in large part, by limitations and inertia that existed in the district and surrounding community. Time and resource constraints and community flux adversely affected the involvement of Raven and Friendship House personnel, and the quality and relevance of services provided in the FSCs. These limitations also narrowed the range of decisions that could be made about goods and services to provide, dictating what could and could not be offered, and how the FSCs would function. Additionally, opportunities for increasing the knowledge of personnel through information sharing and other forms of cooperative activity and staff development seemed enormously hampered by multiple cases, and according to the variability and challenges that the different cases introduced (please see Table 6 Appendix E).

Further, analyses suggest that these factors contributed to those individuals working in support of the FSCs providing assistance to students and families that was passive, not relevant, fragmented, and of questionable value. This became evident when Raven educators complained that while some pressure was exerted by principals to make better use of the FSCs, the teachers and Friendship House employees "really had no time to work together to find out what each other was doing." Teachers indicated that while they "told them about the FSCs," they resisted recommending students because the teachers were uncertain of the changes that were occurring in the centers, and doubtful about the capacity for the centers to adequately assist. Some admitted to "sending [the students] anyway, just to let them see [the FSCs]," and other teachers suggested that "even though [the FSCs] probably couldn't help, you know they're welcome there, and you never know what [the students] might get from going."

Under the school based FSCs described here, analyses revealed that the students' peers, other families, and members of the church and community played a more prominent role in encouraging Raven students and families to obtain assistance. This finding contrasts with other research on school-linked coordinated services that suggests that teachers assume a central role in making referrals (Smrekar, 1994). This finding also seemed to have implications for the rates at which different groups of students and families participate, as well as for the goods and services which they sought after and experienced in the FSCs.

Students, parents and guardians who received greater assistance with clothing, food, child-care, and government assistance, for instance, tended to initiate and experience fewer favorable interactions with Raven administrators and teachers. Instead, their contacts with Raven school personnel were more often initiated by teachers and administrators, suggesting that the centers represented another mode for contacting hard-to-reach families, and that this mode was more successful than written notices or telephone calls for increasing their presence in the schools. In contrast, students and family members who initiated contact, were more visible, and experienced greater academic success in school, also interacted with Friendship House instructors and tutors more often, and were more likely to participate in after and summer school learning. Raven and Friendship personnel offered the familiar refrain, suggesting that "the parents and students you need to see, you rarely get, and the one's who probably don't really need the extra instruction, are first in line to get it."

Regarding goods, services, textbooks and classroom materials available to Raven instructors and tutors, all seven students and each of the 11 parents and guardians concurred that Friendship House "do a good job of getting and giving out things," but that staff did not have adequate school supplies. These students, parents and guardians indicated that they were thankful for the goods and regular assistance they received, and that instructors and staff "counted pencils," "told us to return pencils," and to "be careful with pencils because they didn't have enough." Similarly, knowledgeable students, parents and guardians confirmed that textbooks, workbooks and other materials found in their classroom schools were generally not available at the FSCs. Additionally, when asked to comment about interactions between Raven teachers and Friendship House instructors and staff, the students indicated they believed that the "teachers and instructors and tutors get along good," and "they seem to like each other," and that "sometimes the teachers and instructors know what each other is doing."

Implications for Family Support Centers and Improving Title 1 Schools for Needy Students and Families

This study describes an ongoing struggle between hope and despair. It opens with a discussion about cooperation and teamwork in the form of school-linked social services, and continues with a description of economic and demographic trends in the Raven Elementary Public School District community. Next, a discussion about the connection between the Joseph I. Flores Academia del Pueblo community based facility and six Raven elementary schools is given. This partnering describes the efforts of community members and educators to provide assistance to local students and families through six school-based family support centers. These centers give food, household items and clothing, and provide advice and training to assist individuals and families with their health, housing, budget, and other social welfare concerns, and their efforts at human capital development. After school and summer school instruction is also provided to Raven school students in reading, mathematics, and other subjects. This tutoring is offered to help the students complete their homework, and to increase their chances for achieving academically and finishing school.

Findings and conclusions provided in this study were compiled using qualitative, empirical, and critical techniques. They were organized using prior research on education reform, dropout prevention, and school-linked social services. Specifically, outcomes related to four distinct but interrelated themes were introduced. Each theme describes programmatic efforts that research suggests are needed to achieve change. The findings included in this study describe the degree to which Raven and Friendship House administrators, faculty and staff were successful in working together, providing students and families with needed social welfare and educational support, and addressing the four distinct but interrelated themes. The remainder of this study describes challenges and contradictions embedded in the research that the framework was unable to address.

CHALLENGES AND CONTRADICTIONS TO ACHIEVING MEMBERSHIP

Researchers describe participant membership as concerned with the quality and nature of experiences that students and family members enjoy in and outside of schools. It examines issues related to curriculum, instruction, assessment, school climate, student discipline, and support provided to youth and families in academic and nonacademic areas. Challenges introduced in this study relate to approaches used for assessing student and family progress. Contradictions describe how achieving success is likely to prohibit individuals, schools and communities from experiencing needed change.

Analyses of the findings introduced in this study suggest that the goods and services provided by Raven and Friendship House personnel in the family support centers (FSCs) reflect the offerings of an average school-linked social service collaborative. When low levels of trust among cooperating faculties are combined with school and support center objectives and operations that are only symbolically and intermittently linked, the opportunities that linkages are capable of providing are likely to have limited overall effect.

The cooperating faculties described herein were pretty well divided in their beliefs about providing education and social welfare support. Raven school personnel were geared toward helping students acquire knowledge, academic skills, and positive self-concepts for the future. Friendship House staff braced themselves to address the social welfare needs that students and families experienced in their daily lives.

This division of beliefs was reflected in attitudes and behaviors that witnessed school personnel becoming guarded, doubtful, and resistant to discussing ideas and sharing school supplies. It also led FSC administrators and faculty to grow suspicious of the teachers' commitment to students and families in general, and doubtful about the importance attached by Raven personnel to confronting the students' and families' welfare and social service needs. The contradiction that emerged from this division of responsibilities resulted in Raven and Friendship House personnel wanting in what was best for students and families, and their suspicions and doubts contributing to personnel understanding that given the operation and the nature of the FSCs, Raven students and families were not likely to experience all they were entitled.

CHALLENGES AND CONTRADICTIONS THAT UNDERMINE BELIEF

Discussions about participant beliefs are concerned about the work and professional life of educators and community supporters. They consider the range of new roles and responsibilities that define educators' work. Challenges that affected the capacity for Raven school and Friendship House personnel to increase student and family involvement and the chances for pupils and family members to achieve success, relate to the nature and characteristics of the issues they encountered and to the nature and characteristic of change. Contradictions describe how new initiatives quickly became outdated, and how the promise of involving others in doing more, resulted in the depleting of energy, resources, and less.

Raven's and Friendship House's personnel met student and family needs that were numerous, severe, and overwhelming. Additionally, faculty and administrator turnover, demographic changes, growing impoverishment in the community, and limited money, time, and resources depleted their capacity to provide adequate support. This left Raven and Friendship House personnel very nearly confounded.

Moreover, analyses of the consequences of this constant state of depression and flux suggest that while Raven and Friendship House personnel were committed to providing support, they also perceived that any treatment or service they provided was likely to have limited relevance and long term effect. Contradictions emerged for Raven and Friendship House administrators and faculty in relation to working together and with others in delivering opportunities that might produce desired effects. By working diligently to address the problems and needs of the students and families, in other words, personnel were able to unearth issues that were far more complex, and that were likely to call for and deplete their already tired bodies and meager reserves.

CHALLENGES AND CONTRADICTIONS THAT DEFY GOVERNANCE, MANAGEMENT AND LEADERSHIP

Literature describing governance, management and leadership refers to ways in which accountability and authority are allocated. It also calls for creating new mechanisms for making decisions that involve sharing power with others in the community. Challenges introduced in this study that defied effective practices in these areas relate to the range, severity and unpredictability of student and family needs, and to the inability for individuals to obtain appropriate resources and develop mechanisms effective for delivering coordinated assessment and support. Contradictions to governance, management and leadership relate to questions about freedom and democracy, and to balancing the individual versus the social good.

In their most general and critical sense, governance, management and leadership may be conceived as power, influence, authority, and control. In addition, these terms suggest a capacity for one to do something affecting another. Moreover, it may be argued that the effectiveness of governance, management and leadership may be measured in relation to the capacity for individuals to change a probable pattern of specific events, and in their capacity to move others to ensure that desired objectives are achieved.

In this study, satisfactorily addressing the social and education needs of Raven students and families called for developing close and personal relationships. Additionally, developing responses that were appropriate and effective called for opening one's soul to examination, and for administrators, personnel, and others in positions of authority to bring knowledge, skill, and compassion for working with others to help them achieve resolution.

As with conditions that undermined the confidence and belief that individuals had in their capacity and the capacity of the FSCs to deliver adequate support, the range and severity of the challenges disadvantaged students and families faced, influenced Raven school and Friendship House personnel to experience doubt about their ability to satisfactorily obtain and provide goods, social service, and educational support that were key to ameliorating the social and education needs of Raven students and families. Contradictions emerged for Raven school and Friendship House personnel between their desire to establish kinship with students and families to help them progress on the one hand, and the desire of Raven and Friendship House administrators and faculty to keep the privacy of students and families intact.

CHALLENGES AND CONTRADICTIONS TO SCHOOL AND COMMUNITY DEVELOPMENT

The fourth theme to be found in the literature calls for finding ways for urban schools and communities to improve by enhancing the chances for disadvantaged youth and families to achieve success. It comes from analyses of urban school efforts that involve integrating and coordinating health and social services for children and families, programs for youth employment, incentives and mentoring for higher education, as well as from analyses of research that calls for infusing private sector resources into curriculum and other academic activities that students experience in schools. Challenges that lie ahead and relate to the ideas that follow describe directions for policy makers and future researchers to take. Contradictions arise with the costs and burdens associated with looking ahead, and the expense that accrues from plying inadequate resources and support today for those who are currently at greater risk of falling further behind.

As suggested by analyses conducted for this research, the capacity for school-linked social services to deliver sufficient resources and services may be equivalent to the likelihood for schools and social service providers located in disadvantaged communities to extract support that exceeds the potential for individuals and groups living in these communities to deliver. Indeed, if communities play a role in raising children as so many in our generation suggest, this study makes it obvious that while their affections may very nearly be the same, some communities are better equipped and prepared than others to serve families and youth.

This circumstance suggests a need for more advantaged groups and communities to get involved. It suggests that rather than working together in isolation, agencies like Friendship House and schools like those in the Raven district need individuals in local and state positions responsible for parks and recreation departments, for example, to assume a greater role in taking resources from larger reserves that are likely to result in greater benefits, and directing them toward needy students and families so that they might experience greater opportunities for social and human capital development. Researchers and policy makers may also make a more significant contribution by conducting additional research on school-linked social services, and developing proposals and policies that focus on community to community interaction, and that reward more fortunate individuals and groups for the efforts in bringing the disadvantaged relief.

REFERENCES

Bamgbose, A. (1989). Issues for a model of language planning. *Language Problems and Language Planning, 13,* 24–34.

Bogdan, R. C., & Biklen, S. K. (1992). *Qualitative research for education: An introduction to theory and methods.* Needham Heights, MA: Allyn and Bacon.

Capper, C. A. (1994). We're not housed in an institution, we're housed in the community: Possibilities and consequences of neighborhood-based interagency collaboration. *Educational Administration Quarterly, 30*(3), 257–277.

Capper, C. A. (1997). Critically oriented and postmodern perspectives: Sorting out the differences and applications for practice, *Educational Administration Quarterly, 34*(3), 354–379.

Christian, D. (1988). Language planning: The view from linguistics. In: F. Newmyer (Ed.), *Linguistics: The Cambridge Survey: Vol. 4. The Sociocultural Context.* (pp. 193–209). Cambridge: Cambridge University Press.

Cooper, R. L. (1989). *Language planning and social change.* Cambridge: Cambridge University Press.

Fishman, J. A. (1973). Language modernization and planning in comparison with other types of national modernization and planning. *Language in Society, 2,* 23–42.

Foster, W. P. (1982). *Toward a critical theory of educational administration.* A paper presented at the annual meeting of the American Educational Research Association.

Gardner, S. L. (1993). Key issues in developing school-linked integrated services. *Education and Urban Society, 25*(2), 141–152.

Glaser, B. G., & Strauss, A. L. (1967). *The discovery of grounded theory. Strategies for qualitative research.* Chicago: Aldine.

Glesne, C., & Peshkin, A. (1992). *Becoming qualitative researchers: An introduction.* NY: Longman Publishing Co.

Gordon, I. J., Curran, R. L., & Avila, D. L. (1966). *An inter-disciplinary approach to improving the development of culturally disadvantaged children.* U.S. Department of Health, Education and Welfare, Office of Education, Project No. 5–0698.

Hobbs, N. (1982). *The troubled and troubling child: Reeducation in mental health, education, and human services programs for children and youth.* San Francisco: Jossey Bass, Inc., Publishers.

Hobbs, N., Bartel, N., Dokecki, P. R., Gallagher, J. J., & Reynolds, M. C. (1979). *Exceptional teaching for exceptional learning.* A report to the Ford Foundation. 320 East 43rd Street, New York, N.Y. 10017. Library of Congress Catalog Card Number 79–65207.

Hobbs, N. (1978). Families, schools, communities: An ecosystem for children, *Teachers College Record, 79*(4), 756–766.

Hobbs, N. (1975). *The futures of children: Categories, labels and their consequences.* San Francisco: Jossey Bass, Inc., Publishers.

Hobbs, N. (1966). Helping disturbed children: Psychological and ecological strategies. *American Psychologist, 21*, 1105–1115.

Miles, M. B., & Huberman, A. M. (1994). *Qualitative data analysis: An expanded sourcebook.* Second Edition. Thousand Oaks, CA: SAGE Publications, Inc.

Neustupny, J. V. (1983). Toward a paradigm for language planning. *Language Planning Newsletter, 9*(4), 1–4.

Paulston, C. (1984). Language planning. In: C. Kennedy (Ed.), *Language Planning and Language Education.* (pp. 55–67). London: Allen & Unwin.

Peña, R. A. (1998). *Community based organizations and public schools:* Implications for inclusion, pedagogy and the organizational context of schools. ERIC Document Reproduction Service No. ED 424 649.

Peña, R. A. (1995). Teachers' expectations and students with challenging behaviors. *Current Issues in Middle Level Education, 4*(1), 67–78.

Peña, R. A. (1993). Organizational uncertainty: Public schools and students with challenging behaviors. *Dissertation Abstracts International, No: AAI9330833.*

Sirotnik, K. A., & Oakes, J. (1986). *Critical perspectives on the organization and improvement of schooling.* Boston: Kluwer- Nihhoff Publishing.

Smrekar, C. (1994). The missing link in school-linked social service programs. *Educational Evaluation and Policy Analysis, 16*(4), 422–433.

Smrekar, C. (1993). Rethinking family-school interactions: A prologue to linking schools and social services, *Education and Urban Society, 25*(2), 175–186.

Wehlage, G., Smith, G., & Lipman, P. (1992). Restructuring urban schools: The new futures experience. *American Educational Research Journal, 29*(1), 51–93.

Wehlage, G., Rutter, R. A., Smith, G. A., Lesko, N., & Fernandez, R. R. (1989). *Reducing the risk: Schools as communities of support.* Philadelphia: Falmer Press.

Weinstein, B. (1986). Language planning and interests. In: L. Laforge (Ed.), *Proceedings of the International Colloquium on Language Planning* (pp. 36–58). Quebec: Les Presses de l'Université Laval.

APPENDIX A

Question 1. How would you describe your relationship with students, families Raven school teachers and administrators.

Table 2. Friendship House Response to Questionnaire, Item 1.

Respondent and Numbers of Responses	Relationship with	Question 1. How would you describe your relationship with students, families, Raven school teachers and administrators?				
		Very Good	Good	Average	Not Good	Very Bad
Student	Students	7 (100%)	0	0	0	0
Interviews	Families	7 (100%)	0	0	0	0
	Teachers	5 (71%)	2 (29%)	0	0	0
	Admin.	2 (29%)	4 (57%)	1 (14%)	0	0
Family	Students	11(100%)	0	0	0	0
Interviews	Families	5 (46%)	6 (54%)	0	0	0
(11)	Teachers	7 (64%)	4 (36%)	0	0	0
	Admin.	4 (36%)	7 (54%)	0	0	0
Teacher	Students	12 (86%)	2 (14%)	0	0	0
Interviews	Families	8 (57%)	4 (29%)	2 (14%)	0	0
(14)	Teachers	0	2 (14%)	12 (86%)	0	0
	Admin.	1 (7%)	2 (14%)	11 (79%)	0	0
Admin.	Students	7 (100%)	0	0	0	0
Interviews	Families	7 (100%)	0	0	0	0
(7)	Teachers	0	1 (14%)	5 (72%)	1 (14%)	0
	Admin.	0	1 (14%)	4 (57%)	2 (29%)	0

Question 1. How would you describe your relationship with students, families, teachers and administrators in the FSCs?

Table 3. Responses to Questionnaire item 1 Provided by Raven Personnel, Students and Families.

Respondent and Numbers of Responses	Relationship with	Question 1. How would you describe your relationship with students, families, teachers and administrators in the FSCs?				
		Very Good	Good	Average	Not Good	Very Bad
Students	Students	16 (14%)	76 (68%)	18 (17%)	2 (0.01%)	0
Surveyed	Families	66 (59%)	32 (29%)	14 (0.13%)	0	0
(112)	Teachers	38 (34%)	52 (46%)	16 (14%)	4 (0.04%)	2 (0.02%)
	Admin.	55 (50%)	57 (51%)	0	0	0
Families						
Surveyed	Students	17 (20%)	60 (70%)	9 (10%)	0	0
(86)	Families	21 (24%)	56 (65%)	9 (10%)	0	0
	Teachers	52 (60%)	22 (26%)	12 (14%)	0	0
	Admin.	62 (72%)	11 (13%)	12 (14%)	1 (1%)	0
Teachers	Students	67 (40%)	55 (32%)	31 (18%)	8 (5%)	8 (5%)
Surveyed	Families	47 (29%)	92 (54%)	21 (12%)	9 (5%)	0
(169)	Teachers	7 (4%)	14 (28%)	108 (64%)	36 (21%)	8 (5%)
	Admin.	32 (19%)	53 (31%)	57 (34%)	22 (13%)	5 (3%)
Admin.	Students	4 (24%)	11 (65%)	3 (18%)	0	0
Surveyed	Families	6 (35%)	6 (35%)	5 (30%)	0	0
(17)	Teachers	0	2 (12%)	15 (88%)	0	0
	Admin.	2 (12%)	6 (35%)	8 (47%)	1 (6%)	0

APPENDIX B

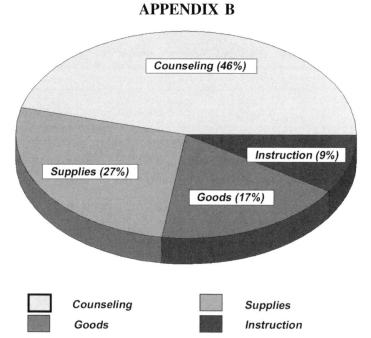

Fig. 3. Use of FSC Service and Goods by Students and Parents.

APPENDIX C

Question 2. How do you rate the goods and services available in the FSCs overall?

Table 4. Participant Responses to Question 2.

Respondent	Number of Responses	Question 2. How do you rate the goods and services available in the FSCs overall?				
		Very Good	Good	Average	Not Good	Very Bad
Students	112	56 (50%)	52 (46%)	2 (0.01%)	2 (.01%)	0
Families	86	44 (51%)	34 (40%)	8 (0.09%)	2 (0.01%)	2 (0.01%)
Teachers	169	5 (0.03%)	14 (0.08%)	74 (44%)	51 (30%)	25 (15%)
Admin.	17	1 (0.06%)	0	11 (65%)	2 (12%)	3 (18%)

APPENDIX D

Question . What can you say about the commitment and caring that others show toward the FSCs?

Table 5. Participant Responses to Question 3.

Respondent	Number of Responses	Question 3. What can you say about the commitment and caring that others show toward the FSCs?				
		Very Good	Good	Average	Not Good	Very Bad
Students	112	44 (39%)	60 (54%)	4 (0.04%)	4 (0.04%)	0
Families	86	33 (40%)	45 (52%)	3 (0.03%)	5 (0.06%)	0
Teachers	169	43 (25%)	56 (33%)	63 (37%)	5 (0.03%)	2 (0.01%)
Admin.	17	6 (35%)	11 (65%)	0	0	0

APPENDIX E

Question 4. What can you say about the level of overall support that is provided to the FSCs?

Table 6. Participant Responses to Question 4.

Respondent	Number of Responses	Question 4. What can you say about the level of overall support that is provided to the FSCs?				
		Very Good	Good	Average	Not Good	Very Bad
Students	112	24 (21%)	60 (54%)	24 (21%)	4 (0.04%)	0
Families	86	18 (21%)	62 (72%)	5 (0.06%)	1 (0.01%)	0
Teachers	169	11 (0.07%)	26 (15%)	60 (36%)	56 (33%)	16 (0.09%)
Admin.	17	0	2 (12%)	12 (71%)	3 (20%)	0

ECONOMIC GEOGRAPHY OF LATINO LOS ANGELES: SCHOOLING AND URBAN TRANSFORMATIONS AT CENTURY'S END*

Rodolfo D. Torres and Louis F. Mirón

INTRODUCTION

There has been a tendency in current debates on Latino education to ignore the emergence of post-Fordist urban transformation in school reform and community development. To understand the connection between Latino schooling and economic restructuring, we must link educational practice to the structural dimensions that shape institutional life – globalization. A first step in this process would entail simply articulating the political and economic contexts of schools. While this may seem abstract, or perhaps even trite, it does not have to be. The central claim of this chapter is that Chicano/Latino education in the Los Angeles Metropolitan area can only be understood if it is seen in relation to changing class and racialized relations. By doing so, we hope to render the conceptual link between Latino schooling and the economic geography of Los Angeles more accessible.

* Portions of this chapter have appeared in Latino Metropolis, Victor Valle and Rodolfo D. Torres, University of Minesota Press, 2000.

Community Development and School Reform, Volume 5, pages 101–120.
ISBN: 0-7623-0779-X

URBAN TRANSFORMATIONS

The closing years of the twentieth century represent the culmination of major changes in the socioeconomic landscape of U.S. society – changes that potentially could herald greater class conflict and social unrest than U.S. modern history has ever known. And nowhere are these changes more evident than in urban centers, where large populations have been directly affected by the impact of economic restructuring, postindustrial conditions of urban life and the globalization of the economy.

In discussing urban transformations, we begin with the precipitous growth of the immigrant and refugee populations in urban centers of the United States. Since the late 1960s, immigration patterns to the United States have changed in at least two important ways.[1] First, there has been a significant increase in the overall annual entry levels: from about 400,000 a year in the 1970s to 600,000 a year in the 1980s to about 800,000 a year in the 1990s (Westphal, 1998, p. 13). As a consequence, the foreign-born population of the U.S. now stands at over 25 million, which means that roughly one out of every ten residents was born abroad, making it the nation's highest proportion since before World War II (Westphal, 1998). Second, there has been a marked shift in the regional composition of migration flows. In 1940, approximately 70% of all immigrants entering the United States came from Europe, while in 1992 most of them came from either Asia (37%) or Latin American/Caribbean (44%), Europe accounting for only 15% of the immigration pie (The numbers game, 1994, p. 15).

Given these two changes, the U.S. has become increasingly racially and ethnically diversified. This effect has been particularly profound for those states in which the latest waves of immigrants tend to cluster, these being California, New York, Illinois, New Jersey, Florida and Texas, which together receive about three-quarters of all new immigrants (Sassen, 1989). And it has been even more true for certain large metropolitan areas, such as New York, Los Angeles, San Francisco, Chicago, Houston and Miami, where immigrants and their children make up a notably higher proportion of the population than they do of the U.S. total (Sassen, 1989). Many U.S. cities have thus been transformed from primarily European-American metropolises into meeting places for a wide rage of peoples from all over the world, into zones that showcase the juxtaposition of different societies, modes of life, and social practices.

This demographic transformation has been most marked in California, which has the highest percentage of foreign-born residents of any state in the nation. It is estimated that about one-fourth of the state's 33 million inhabitants was born outside of the U.S. (about 2 million of these foreigners are undocumented immigrants). California also leads all other states in the number of new legal immigrants that settle within its borders each year, which is calculated to be anywhere from 200,000 to 300,000 (Garcia y Griego & Martin, in press). In addition, it is estimated that roughly 105,000 Mexicans have settled illegally in the U.S. each year since 1990 (del Olmo, 1997), a good number of these undoubtedly in California. Some researchers suggest that the immigrant population of California will only continue to grow into the next century (Garcia y Griego & Martin, in press).

New Immigrants in a Changing Economy

Earlier waves of immigration from Asia and Latin America (then primarily Mexico) notwithstanding, the "new immigrants" are entering a society that is vastly different from that entered by their predecessors. For one, the high-wage manufacturing jobs that were once the basis of a largely middle-class society have been exported overseas, having been supplanted by skilled professions in the information economy that require specialized training through years of increasingly costly education at the postsecondary level.[2] At the lower end of the service and information economy are the legions of Asian and Latino laborers who hold ethnically typed low-wage jobs cleaning, clothing, feeding, and housing those on the other side of the widening class divide. It is the janitorial, clothing, agriculture, and construction industries that are the principal employers of immigrant workers. In extreme instances, immigrants have been found to be working under conditions comparable to slavery.[3]

In the past thirty years, Latinos have become an increasingly important segment of the immigrant population, one whose growing presence and conditions are closely intertwined with the very forces that are causing the ongoing economic restructuring and reshaping of once-familiar international, national, regional, and local landscapes. For nearly four decades, these global political economic changes have greatly stimulated Latino immigration to the United States.

On March 1, 1999, there were an estimated 31.7 million Latinos in the United States, constituting 11.7% of the total population. By 2005, the Census Bureau projects that Latinos will overtake African Americans as the largest minority group in the United States. By 2050, there should be more Latinos than all other minorities combined. As of 1998, the Latino population in California stood at 7,687,938, or 26% of the state's total population; California could have a Latino majority by the year 2040, according to recent Census Bureau projections. The Latino population of Los Angeles County in 1998 was 4,226,000, or 44%; in the city of Los Angeles, where the Latino population was nearly 1,651,000, Latinos made up 45% of the total population.[4]

Perhaps more important than sheer numbers, Latinos are the most significant and fastest growing sector of the working class in the United States. Within a few years, Latinos will make up more than a quarter of the nation's total workforce, a proportion that is more than three times larger than this group's proportion of the total population. Equally significant, Latino men and women are increasingly concentrated in the very industries that have been most influenced by the economic restructuring of the United States. J. Scott & A. S. Paul (1991) have described Southern California as the most advanced case of post-Fordist industrialization in the United States, if not the world.[5] Los Angeles County, for example, is the "postindustrial" heartland of the United States, with its industries based on non-unionized low-wage workers who are drawn increasingly from the ranks of immigrants, legal and undocumented. The county is the nation's largest manufacturing center (Kyser, 1999, p. 2), with 667,800 workers employed in manufacturing as of 1998.[6] It has been estimated that half of these workers are Latinos.

In metropolitan Los Angeles, the economy has increasingly come to rest on this cheap, non-unionized labor. The deindustrialization of Central Los Angeles in the recent past is linked directly to the reindustrialization of the Eastside based upon low-wage manufacturing. (Valle & Torres; 1993, 1994) Here Los Angeles seems unique among global cities (see Sassen, 1991). These two interconnected transformations – the spread of the post-Fordist industrial development and the accompanying cycles of deindustrialization of the Los Angeles urban core – were greatly shaped by capital's virtual monopoly over Los Angeles County land use policies.

The Polarization of Los Angeles

In Los Angeles the rich are getting richer and the middle class is besieged by the threat of unemployment and rising debt levels (Allen & Turner, 1997). A report released in 1998 by the California Assembly Select Committee on the California Middle Class, chaired by Assemblyman Wally Knox, indicated that income inequality in Los Angeles has increased significantly. The study on which the report was based found that as of 1996, 41% of the residents of L.A. County lived in households with annual incomes below $20,000, and fully two-thirds lived in households with annual incomes below $40,000. Only 26% were in middle-income households making between $40,000 and $100,000, with 8% in households making more than $100,000.

California's recovery from the recession of the early 1990s has not mitigated this trend, but rather has magnified the effect of structural inequalities in L.A.'s economy. According to an analysis undertaken by the *Los Angeles Times* in

1999, nearly all the job growth in Los Angeles County, since the low point of the recession in winter 1993, has been in low-income jobs (Lee, 1999, p. 1). Although the number of new jobs created is impressive, almost three hundred thousand since 1993, very few of these jobs fall in the middle-class income range of $40,000 to $60,000. L.A.'s economic recovery has produced far more parking lot attendants, waiters, and video store clerks than highly paid workers in information technology, entertainment, or international trade. The majority of new jobs pay less than $25,000 per year and barely one new job in ten averages $60,000 per year. Many new jobs also lack long-term security or health care benefits.

The impact of these low salary figures is even more dramatic in light of the elevated cost of living in Los Angeles County. The high cost of real estate in Los Angeles makes it difficult for low-income workers to buy homes even if several wage earners share the same household. Whereas neighboring Orange County has seen a 10% increase in its homeownership rate in the past ten years, the rate for Los Angeles County has scarcely moved in the same period. According to Mark Drayse, research director of the nonprofit Economic Roundtable, the net effect of low salaries and high living cost is that the population is "becoming more polarized" (Lee, 1999, p. 1).

The current socioeconomic condition of Latinos in Los Angeles can be traced directly to the relentless emergence of the global economy and recent economic policies of expansion. Policies such as the North American Free Trade Agreement (NAFTA) have weakened the labor participation of Latinos through the transfer of historically well-paying manufacturing jobs to Mexico and other "cheap-labor" manufacturing centers around the world. Such consequences highlight the need for scholars to link the condition of U.S. Latinos in cities to the globalization of the economy. Few scholars have contributed more to our understanding of globalization and economic restructuring than Saskia Sassen, who posits:

> Trends in major cities cannot be understood in isolation of fundamental changes in the broader organization of advanced economies. The combination of economic, political, and technical forces that has contributed to the decline of mass production as the central element in the economy brought about a decline in a wider institutional framework that shaped the employment relations (Sassen, 1996).

In light of this view, Latino Los Angeles can be fully understood only within the context of the U.S. political economy and the new international division of labor. Without question, the United States is the wealthiest country in the world, yet it is the nation-state with the greatest economic inequality between the rich and the poor and with the most disproportionate wealth distribution of all the "developed" nations of the world. To overlook this economic reality in the analysis of Latino populations is to ignore the most compelling social phenomenon in U.S. society today – the growing gap between rich and poor (Freeman, 1999).

THE POLITICAL ECONOMY OF LANGUAGE

Historically, the relations between the institution of public schooling and cities generally have been characterized as isomorphic owing to the divorce of "politics" from "education." Of course, no such split can be realized, and despite the apparent good intentions of education reformers in the progressive and subsequent eras, the practices of education are both political and politicized (see Mirón, 1997).

On this note, California's Proposition 227 ("English for the Children"), a ballot measure passed by the voters in 1998, can be more fully understood in light of a symbolic statement of fear, hostility, and frustration emanating from the inability to effect the transformations taking place in the United States. During this period, the political and cultural climate for immigrants has been distinctly hostile, and there have been two interrelated foci of anti-immigrant sentiment: one targets the economic consequences of immigration, the other its cultural effects. The first anti-foreign sentiment targets immigrants, principally the undocumented, for their utilization of welfare, education and health care services. It is primarily embodied in California's Proposition 187, which was passed by the voters of California on November 8, 1994.[7] The prevalent belief at the time was that immigrants cost California millions of dollars a year in public services, depleting the state of scarce financial resources, and precipitating a decline in the general "health" of the state. Reductions in programs for disease control were seen as leading to the spread of diseases. Overcrowded classrooms, increased class size, and reduced teacher-to-student ratio were perceived as leading to a deterioration of the quality of education, thus reducing California's competitiveness in the global economy (Citizens for Legal Immigration Reform/Save Our State, n.d.). So the hope was that the denial of public benefits would reduce, if not altogether stop, the flow of undocumented immigrants, as well as force those who were already inside the state to leave the country. Thus the state would be relieved of a major burden on its coffers, paving the way for a more prosperous California.

The second locus of anti-immigrant sentiment targets the detrimental cultural consequences of immigration. This focus is most visible in the current antipathy towards the use of non-English languages; much of it derived from the fear that linguistic difference will fragment the American nation.[8] Or, to be more precise, the antipathy is the product of five basic assumptions about language (and here we borrow from James Crawford's (1992) characterization of the English-only movements of the 1980s): (1) "English has been Americans' strongest common bond, the 'social glue' that holds the nation together;" (2) "Linguistic diversity inevitably leads to political disunity;" (3) "State-sponsored bilingual services remove incentives to learning English and keep immigrants out of the mainstream;" (4) "The hegemony of English in the United States is threatened by swelling populations of minority-language speakers;" and (5) "Ethnic conflicts will ensue unless strong measures are taken to reinforce 'unilingualism'" (p. 24). So, as George Sanchez (1997) has noted,

> Despite the fact that English has become the premier international language of commerce and communication, fueled by forces as widespread as multinational corporations, the Internet, popular culture and returning migrants, Americans themselves consistently worry that immigrants refuse to learn English and intend to undermine the pre-eminence of that language within American borders (p. 1020).

But it is really not all immigrants who are targets, it is primarily Latino immigrants; and it is really not all languages that are suspect, it is principally Spanish. According to Thomas Muller (1997), "The growth of Spanish has now reached a level that the nation is close to being a two-language society, causing . . . uneasiness among those who believe the English language should have exclusive domain" (pp. 109–110). Perhaps more worrisome is that this emerging two-language society seems to have the tacit recognition not only of the business community, which employ Spanish in the media, popular culture and advertising, but of the government as well, which provides bilingual education to students whose mother tongue is not English. More specifically, these sentiments have been channeled into Proposition 227, which seeks to end bilingual education.[9] In a way, then, one of the main purposes of Proposition 227 is to weaken the entrenchment of Spanish in American society. As such, it promises to remove one of the obstacles preventing the proper functioning of the "traditional process[es] whereby newcomers to America and their children gradually and voluntarily assimilate into our common American culture" (Taylor, 1998, World Wide Web).

A basic tenet held by movements exemplified by Proposition 227 is the need for everyone residing in the United States to learn English. In a study administered in Santa Ana,[10] Mirón, Inda & Aguirre found that though English learners recognized a need to maintain their use of Spanish, at the same time they often place a lot of faith on English as a mechanism for economic betterment. They talk about English as "being the language of economic opportunity," one that allows immigrants "to fully participate in the American Dream of economic and social advancement" (Unz & Tuchman, 1998). Yet they also recognize the equally important need to maintain their use of Spanish, both for cultural and familial reasons, and to give them an advantage in the marketplace (see Mirón, Inda & Aguirre, 1998).

THE PROCESS OF CLASS

The globalization of capital and its changes in class relations form the very backdrop of contemporary Latino schooling, politics, and racialized relations, but is conspicuously absent in most contemporary accounts of Latino life in the United States – accounts that ignore the increasing significance of class and the specificity of capitalism as a system of social and political relations of power.

One issue is clear to us. Despite claims by some on the Left (whose theoretical orientations could be described as "postmodernist" or "post-Marxist") that class politics is an anachronism, we maintain that the concept of class has increasing analytic value as we enter the twenty-first century. We are guided here by David Harvey's (1996) definition of class:

> I insist that class is not a thing, an entity, or a "permanence" (though under given conditions it can indeed assume such a form) but fundamentally a process. But what kind of process? Marx appears to define class relationally as command (or non-command) over the means of production. We prefer to define class as situatedness or positionality in relation to process of capital accumulation (p. 359).

We also recognize that there is a theoretical tension between our insistence on the need for a renewed class approach at a time when it has become fashionable for some on the academic Left to question its analytic utility, and our often implicit constructivist and discursive mapping of racialized relations and identities in Los Angeles. We argue, nonetheless, that a political economy approach informed by both a Marxist and a critical "postmodern" social theory offers the best way to theorize about Latinos in Los Angeles in the context of demographic shifts, changing class formations, and new forms of "global" capitalism. In a recent interview, Stuart Hall echoes our concern with the Left's silence on this issue of class and its failure to articulate sufficiently the relationship between the economic and the political in discussions of global capitalism:

> I do think that's work that urgently needs to be done. The moment you talk about globalization, you are obliged to talk about the internationalization of capital in its late modern form, the shifts that are going on in modern capitalism, post-Fordism, etc. So, those terms which were excluded from cultural studies . . . now need to be reintegrated. . . . In fact, I am sure we will return to the fundamental category of "capital." The difficulties lie in reconceptualizing class. Marx it seems to me now, was more accurate about "capitalism" than he was about class. It's the articulation between the economic and the political in Marxist class theory that has collapsed (Chen, 1996, p. 400).

AFTER "RACE" IN THE METROPOLIS

While the relationship between globalization of capital and concurrent changes in class relations is glaringly absent, the problematic concepts of "race" and "ethnicity" have long been central concepts in sociological discourse and public debate. Policy pundits, journalists, and conservative and liberal academics alike work within categories of "race" and "ethnicity" as though there is unanimity as to their analytic value. Racialized group conflicts are framed and advanced as "race relations" problems, and are presented to the public mostly in black/white terms.

Founded upon this image is what we call the zero-sum picture of the great melodrama of "race relations" in Los Angeles: "racial" groups are considered to be deeply at odds with each other, each group "naturally" apart from others and antagonistic toward members of other groups. Benefits to one group are – or are perceived to be – costs to another. This "race relations" paradigm marks the racialized divide as much between those presumed "black" and "brown" as between those considered "black" and "white," or "white" and "brown."

Academics, media reporters, and politicians "conspired" to use the vocabulary of race to make sense of the 1992 Los Angeles riots because it is a central component of everyday commonsense discourse. And when it became overwhelmingly apparent that it was not a "black/white" riot, the language of "race" was nevertheless unthinkingly retained by means of a switch to the use of the notion of multiracial to encompass the diversity of historical and cultural origins of the participants and victims. Thus, although the "race relations" paradigm was dealt a serious blow by the reality of the riots, the vocabulary of race was retained. But, and here we find the source of the problem, the idea of race is so firmly embedded in commonsense that it cannot easily encompass a reference to Koreans or Hispanics or Latinos, for these are neither "black" nor "white." It is thus not surprising that pundits and scholars stumble over racial ambiguity. The clash of racialized language with a changing political economy presents challenges for scholars and activists alike.

If one had begun with an analysis grounded simultaneously in history and political economy rather than with the supremely ideological notion of race relations, one would have quickly concluded that the actors in any riot in central Los Angeles would probably be ethnically diverse. Large-scale inward migration from Mexico and Central America and from Southeast Asia into California has coincided with a restructuring of the California economy. As a consequence of the loss of major manufacturing jobs and large-scale internal migration within the urban sprawl of greater Los Angeles, the spatial, ethnic, and class structure that underlay the Watts riots of 1965 had been transformed into a much more complex set of relationships.[11]

The notion of "race" and the related concept of "race relations" in social theory and contemporary urban social relations must be problematized. The ideas of race and race relations have been questioned analytically for more than a decade within European academic discussion, and it is only recently that some U.S. scholars have begun to consider the rationale and implications of that critique.[12]

In our analysis of racialized urban relations, we advocate expanding the contemporary sociological debate by arguing for a complete rejection of the use of the terms *race* and *race relations* in academic and public discourse. Perhaps a conceptual framework of "racialization" may be introduced as an alternative model to the sociology of "race relations." Following Robert Miles' seminal work:

> racialization refers to those instances where social relations between people have been structured by the signification of human biological characteristics in such a way as to define and construct differentiated social collectivities . . . The concept therefore refers to a process of categorization, a representational process of defining an Other (usually, but not exclusively) somatically (Miles, 1989, p. 75).

Furthermore, we add that the process of categorization is not only structured around biological characteristics but also around language. Thus racialization can be identified in the "English for the Children movement" as English learners are repeatedly defined as the Other, in terms of their need to acculturate to the dominant norms of the United States. Cultural identity and notions of ethnicity are partly politically formed, rather than embedded in the color of the skin, mother language, or a given nature (Hall, 1990). Hence it is impossible to comprehend the social construction of Latino identities and the impact of schooling upon Latino students without critically addressing the context of racialized relations that gives rise to public education in the United States.

By recognizing the racialization process as the underlying factor in social relations, we can understand the process of signification of one group by another in racialized struggles and tensions. This is important not only in the context of social theory, but also in the context of contemporary politics and school reform. This reexamination, stripped of the "race" language, reveals the role of ethnicity and ethnic politics in shaping the discourse on racialized turf wars in Los Angeles. But there's one big problem. The parochial politics of turf-claiming and displacement won't turn Los Angeles into a world-class city or boost its chances of being the capital of the Pacific Rim. At a time when the city desperately needs unifying influences, the politics of "racial" division serves only to obscure growing racialized class inequality. Ethnicity is substituted for vision and loyalty for leadership.

No one can deny that Los Angeles needs Latino leadership in business, government, the community and education. The Los Angeles Unified School

District needs it most of all: a visible symbol, a cultural voice and an advocate for the families whose children are more than half the student population. But picking a Latino leader won't solve the district's problems, just as picking African American superintendents didn't solve problems in a score of eastern cities. For Los Angeles to have successful Latino school leadership, the city needs a transformational politics that views schools as the driving force in rebuilding Los Angeles as a place to live and to prosper in the global economy.

In analyzing these new social and racialized relations, we posit, there is no need to employ the concept of "race." Indeed, its retention is a significant hindrance in social theory, politics, and education policy. We posit further that class is far more important than the specious concept of race in determining the life chances of Latinos in Los Angeles. However, we do not reject the concept of racism. We maintain that it is necessary for analysts to draw upon the concept of racialization as a tool that will enable them to grasp and map the changing contours of racism(s) and the broad array of structural economic and political inequalities. Through its critical questioning of mainstream assumptions, this analytic approach may allow us to theorize about ethnicity, culture, and racialized social and spatial relations in the contemporary metropolis.

RETHINKING THE FOUNDATIONS OF RESEARCH ON LATINOS, THE ECONOMY, AND EDUCATION

Aligned with oversimplified notions of race and culture, much of the theoretical discourse related to Latinos and schooling has been founded on myopic traditional perspectives which have engaged the Latino population in the United States as a monolithic entity over the past thirty years. These discussions have oftentimes revolved around issues of cultural and linguistic difference. The consequence has been to perpetuate static notions of culture. Such ahistorical and apolitical discussions have generally failed to link notions of culture with a structural analysis of socioeconomic conditions in the United States.

Much of the study of Latino populations emerged out of conditions that have been described as academic colonialism – conditions for legitimation which required studies to be formulated along the very traditional social science values and methods which generated many of the problems faced by Latinos. Chicano/Latino studies programs consistently challenged this academic colonialism. These programs called for a new paradigm for academic scholarship in the field that addressed the problems inherent in standards of legitimation and questioned parameters defined by the academic enterprise in general.

One of the central issues in the struggle to reconstruct the foundations of research approaches to the study of Latinos was the need for a new language to describe the phenomenon of subordinate groups. What this coming to a new language implies is a breaking away from disciplinary and racialized categories of difference and structural inequality. For example, if we consider the literature of the civil rights era and the era of multiculturalism, what is consistently reflected are deeply racialized discourses grounded in black and white categories. This narrow framing has perpetuated a dimension of invisibility with respect to the role of Latino scholarship in larger debates about educational theory and practice.

In addressing the need for a new language there are specific elements that this discourse encompasses. First of all, it is a discourse that is recognized as both simultaneously contextual and contested, and that challenges static and essentialized notions of culture, identity, and language. Secondly, it is rooted in the centrality of the political economy as a significant foundation for understanding how issues of cultural change and ethnicity intersect with the broader structural imperatives of late capitalism. Thirdly, it calls for a rethinking of categories such as "race" and "ethnicity" with respect to the manner in which these can either function to obstruct or further the political project for a cultural and economic democracy in this country. And lastly, such a discourse argues for the defining of a "working canon" of Latino education that is grounded in a critical discourse which avoids the analytical pitfalls of traditional discourses of multiculturalism and cultural studies of the past.

Dialectics of Landscape: Toward a Language of Cities and Work

The concept of "landscape" is borrowed from Sharon Zukin's writings (1991). The novelty of her work stems from the way she expands the meaning of landscape beyond a strictly geographical definition. For her, human landscapes are constructed from the dynamic linkages among a totality of socio-economic, cultural, and political spaces. Existing social and political institutions and the changing material conditions of the marketplace, she argues, shape a landscape's spatial dimensions:

> In a narrow sense, landscape represents the architecture of social class, gender, and "race relations" imposed by powerful institutions. In a broader sense, however, it connotes the entire panorama that we see: both the landscape of the powerful – cathedrals, factories, and skyscrapers – and the subordinate, resistant, or expressive vernacular of the powerless – village chapels, shantytowns, and tenements (Zukin, p. 16).

Zukin's multidimensional approach to landscape reading provides useful analytic categories for describing the architecture, organization and loci of power in Latino Los Angeles. Her approach also provides a method of narrating the Eastside's construction in time by focusing upon the dialectics of capital accumulation and the formation of local political institutions. Explicit in this economic-political dialectic is our representation of the Latino Eastside as a contested terrain where its actors – indigenous or outside capital, indigenous social classes, and local political bureaucracies – vie for strategic advantages.

The spatial convergence of capital, politics and class, Zukin argues, can be portrayed as a dialectic of places. The marketplace, the industrial park, and the suburban neighborhood are defined by their distinct roles as places of production or consumption. According to Zukin, these roles take on a revealing character in the transition to a post-industrial society: "Those places that remain part of a production economy, where men and women produce a physical product for a living, are losers. To the extent they do survive in a service economy, they lack income and prestige, and owe their souls to bankers and politicians. By contrast, those places that thrive are connected to real estate development, financial exchanges, entertainment – the business of moving money and people – where consumer pleasures hide the reins of concentrated economic control" (Zukin, p. 14).

The dialectics of place acquire distinct spatial expressions in late 20th Century, post-industrial cities. Zukin and others have shown that the spatial expansion of urban landscapes proceed along centrifugal force lines, perpetually moving outward beyond the fringes of previous periods of industrial and suburban development. Eventually, the older urban landscape is left behind to atrophy as each succeeding growth ring moves the benefits of capital further from the original urban core. And, as labor follows capital's migration, new cities coalesce to serve new landscapes of production and residence.

At first reading, Zukin's dialectic of places appears to disadvantage Latino barrios since they are often depicted in popular and academic literature as repositories of marginalized surplus labor. Though applicable to certain East Coast Latino barrios, and parts of South and Central Los Angeles, such broad generalizations overlook the Eastside's peculiarities and strategic advantages.

Therefore, rather than place it in the context of an imploding inner city, the Eastside's spatial expansion flows directly from the centripetal economic forces. In other words, the Eastside's shape and size are a direct consequence of its transformation from a collection of semi-rural "edge suburbs", to a still-flaring arch of newer and maturing cities and suburbs.

Seen from the ground, the Greater Eastside appears like a random patchwork of uneven developments marked off by freeways, concrete-lined rivers, rail lines, and landfills, the Hopewell-like monuments of a throw-away consumer civilization. However, seen from the air, or on the demographer's map, the landscape's seemingly convoluted patterning reveals its elongated multi-celled structure. Socially and economically differentiated Latino suburban cells, manufacturing zones, and commercial districts grow eastward from the organism's nucleus, which is just beyond the city's old urban core. Arterial freeways link the old core to the expanding edge, facilitating the circulation of goods, people, and information. At the organism's amorphous edges, privileged suburban cells encroach upon places created by older forms of capital accumulation. The newer suburbs wedded to globalized capital are rewarded with more hospitable landscapes, while those wedded to obsolete modes of production lose control of the ability to recreate their neighborhoods.

The Greater Eastside was, and continues to be, shaped by the destructive and creative energies unleashed in the competition between older and newer forms of capital accumulation. Whether post-industrial transformation is seen as the result of a shift toward a service economy, or as the most recent expression of the continued economic supremacy of industrial manufacturing, is no concern. The focus is on providing a modest rendering of the conceptual link between Latino schooling and the micro economic geography of Los Angeles.

Understanding Latino Education Within a Larger Context

There is a crucial link between economic changes in Los Angeles and in this country, and the economic restructuring that is occurring worldwide. Economic structures are evolving and changing from a Fordist to a post-Fordist or "post-industrial" organization of capital accumulation, distribution, and labor processes. These changes call for new efforts to understand the new class structures of "post-industrial" societies and the changing processes of social stratification and mobility. Needless to say, educational policy considerations are central to the above project as they relate to rethinking the deepening globalization of production, the breakup of working class communities, and the limits and contradictions of state intervention in late twentieth-century capitalism.

While we cannot ignore that the future of schools will be conditioned by social and economic changes, it is by no means predetermined by those changes. A political and ideological battlefield surrounding the role of schools in the changing economy remains. Researchers who are seeking to discover ways to effectively improve the educational conditions of Latino students cannot afford to shy away from entering into this murky realm of contestation that gives shape to the terrain of American public schooling.

CONCLUDING REMARKS

We maintain that the economic forces that have transformed Los Angeles into one of the nation's most dynamic industrial landscapes require a rethinking of role of schooling in Latino political and economic life. The absence of an analysis of class relations with its structural inequalities of income and power represents a serious shortcoming of contemporary Latino education research. Future policy research on Latino education, and political and economic conditions, must be recast in a more rigorous analytical and theoretical framework. To date, research on the political economy of Latino communities has been mired in the unquestioned categories of "race relations." As a result, preconceived notions of what constitutes a Latino community have prevented many from perceiving the actual conditions of Latino education and its relations to the political and economic landscape. Such a lack of theoretical imagination has, in turn, led many to reproduce outdated categories of political analysis on subjects such as bilingual education and post-riot economic development. Unable to advance an independent policy discourse, Latino scholars and community and elected leaders, have been drawn into no-win debates with African American communities on the distribution of economic revitalization funds.

An analysis of the dialectics of landscape can help reveal the locations of actual and potential political power. Arriving at this understanding requires mapping emerging forms of new technologies and identifying the changing organization of work and politics of education in "post-industrial" Los Angeles. As Latino leaders and education activists contest public space in a changing city, they must fight concrete political battles and also "deconstruct" the elite discourses that promote top-down corporate economic development. In other words, because the symbolic landscapes and political economy reinforce each other, political struggles must engage both at once. For community members, school leaders, and educators, performing this task should yield possibilities of democratic economic reform and meaningful change in public schools.

But a system of schools requires very strong grass-roots leadership, rather than the sort of fealty to the hierarchy that old-style bureaucracies reward. And here is where Latino leadership weighs in heavily, not with a single "Latino leader," but with the careful nurturing of a generation of leaders. Unless today's Latino leadership solves the problem of expansion of its ranks and collaboration with other ethnic groups, it creates no place for itself except an endless brown-against-black struggle that further estranges the rising middle classes of all groups and reinforces the racialized and class divide in the city. It fails to address the creation of democratic schools where the population is of shifting ethnicity and where children will live and work in a different economy than their parents' world. Yes, Los Angeles schools need Latino leadership, but more than finding a single leader, the city needs to create the conditions for success in school transformation. Whether such possibilities can be realized remains to be seen, but the benefits could be enormous – a more democratic economy and prosperous Latino community.

NOTES

1. These demographic transformations have a lot to do with the changing nature of the U.S. economy. As industrial production has moved oversees, to take advantage of wage differentials, the traditional U.S. manufacturing base has deteriorated and been partly replaced by a downgraded manufacturing sector, one characterized by an increasing supply of poorly paid, semi-skilled or unskilled production jobs. The economy has also become more service-oriented. Financial and other specialized service firms have thus replaced manufacturing as the leading economic sectors. This new core economic base of highly specialized services has tended to polarize labor demand into high-skill and low-skill categories. The upshot of all this, then, is that these changes in the economy, particularly the creation of low-skill jobs, have created the conditions for the absorption of vast numbers of workers. For a longer exposition of these economic transformations see Saskia Sassen (1991, 1996).

2. As Jeremy Rifkin (1995) observes, white-collar jobs also are being eliminated at a rapid rate – more than three million during the past ten years, with more to follow. Although Rifkin does not specifically address the issue, the continuing displacement of jobs through technological innovation will only serve to heighten hostility toward Asian and Latino immigrants at all levels within the economy.

3. In the summer of 1995, a garment subcontractor in El Monte, California, was discovered to have been holding seventy-two Thai immigrants – mostly female – in virtual bondage. See George White, "Workers Held in Near-Slavery, Officials Say," The Los Angeles Times, August 3, 1995, A1, A20. Despite having one of the more robust economies in Southeast Asia, Thailand has developed a thriving trade in illegal immigration to the United States and other advanced capitalist societies. See John-Thor Dahlburg, "Smuggling People to U.S. Is Big Business in Thailand," The Los Angeles Times, September 5, 1995, A1, A8-A9.

4. County of Los Angeles, 1998, Urban Research Division, Chief Administrative Office, unpublished report released October 5, 1998. In the Los Angeles Unified School District, as of 1998, Latinos made up 69% of the total enrollment of 655,889. See Los Angeles Unified School District, Information Technology Division, "Ethnic Survey Report: Fall 1998," publication 131 (Los Angeles, December 1998). During the 1997-98 school year, more than half of the students in L.A. County public schools were Latino (57% of the total enrollment of more than 1.5 million). See United Way of Greater Los Angeles, State of the County Report: Los Angeles, 1998-99 (Los Angeles, March 1999). New census estimates show that Latino populations are increasing in other Southern California counties. Orange County is estimated to have 760,000 Latinos, making up 28% of the total county population, the fifth highest in the nation. See Megan Garvey,

"Power in Numbers," Los Angeles Times, September 14, 1999, B1. The number of Latinos is expected to increase to 1.9 million, or 48% of the population of Orange County, according to projections released December 17, 1998, by the California Finance Department. See David Haldane, "O.C. in 2040: Near Majority of Latinos, Far Fewer Whites," Los Angeles Times, December 18, 1998, B1.

5. Though economic changes lie at its core, "post-Fordism" represents, according to several leading theorists, a wide-range of social, political, and cultural forms. We recognize that there is considerable theoretical debate over the changing nature and direction of the modern capitalist economy (see Amin, 1994; Wood, 1996; Aglietta, 1998). There are competing opinions about the extent and meaning of these changes and whether they represent a new kind of epochal shift in the basic logic of capitalist accumulation. In the age of "postmodern" excess where critiques of capitalism seem to be out of fashion, we maintain that class and capital still matter. For a perceptive and critical reading of the postmodern project, see Democracy Against Capitalism: Renewing Historical Materialism, by Ellen Meiksins Wood. We find much to admire and learn from Wood's timely critique of postmodernism and its failure to treat capitalism with analytical specificity. Nevertheless, our use of a critical "post-Fordist" framework to theorize about Latino Los Angeles is firmly rooted in Marx's critique of political economy.

6. In 1998, Los Angeles County, with its 667,800 workers, was able to remain the nation's largest manufacturing area for the second consecutive year.

7. Although the measure was overwhelmingly approved (59% to 41%) by the voters of California, it never went into effect. Its main provisions were declared unconstitutional.

8. Although Proposition 187 highlighted the economic costs of immigration, cultural factors were just as crucial, although more understated (see Buchanan, 1994).

9. Of course, support for Proposition 227 is not strictly limited to the "English for the Children" movement. There is good cause for concern among Latino parents, that, if not probably used, bilingual education could leave students in an unenviable position: neither proficient in Spanish nor in English.

10. Santa Ana is located in Orange County, south of Los Angeles. Though the findings in this study are by no means generalizable, we believe they may nonetheless apply to certain highly populated, low-income areas of Los Angeles, namely in East L.A.

11. For an impressive study of the Watts uprising and an informative epilogue on the 1992 civil unrest see, Gerald Horne, Fire this Time: The Watts Uprising and the 1960s.

12. For relevant discussions of race and race relations, see Miles (1989, 1993), Guillaumin (1995), Wieviorka (1995). For implications of the critiques, see Goldberg (1993), Small (1999), Torres and Ngin (1995), Valle and Torres (1995), Darder and Torres (1998), McLaren and Torres (1999), Appiah (1996), and Fields (1990).

ACKNOWLEDGMENT

We would like to acknowledge the competent editorial assistance of Annalisa Mirón.

REFERENCES

Aglietta, M. (1998, November/December). Capitalism at the turn of the century: Regulation theory and the challenge of social change. *New Left Review*, 232.

Allen, J. P., & Turner, E. (1997). The ethnic quilt: Population diversity in Southern California. Northridge, CA: California State University, Northridge, The Center for Geographical Studies.

Amin, A. (1994). *Post Fordism: A reader*. Oxford: Blackwell.

Appiah, K. A. (1996). Race, culture, identity: Misunderstood connections. In: K. A. Appiah & A. Gutmann (Eds), *Color conscious: The Political Morality of Race*. Princeton, N.J.: Princeton University Press.

Buchanan, P. (1994, October 23). What will America be in 2050? *The Los Angeles Times*, p. B7.

California Assembly Committee on the California Middle Class. (1998, May 16). *The distribution of income in California and Los Angeles: A look at recent current population survey and state taxpayer data*.

Chen, K. (1996). Cultural studies and the politics of internationalization: An interview with Stuart Hall. In: D. Morley & K. Chen (Eds), *Stuart Hall: Critical Dialogues in Cultural Studies*. London: Routledge.

Citizens for Legal Immigration/Save Our State. (n.d.). *Proposition 187: The "save our state" initiative: The questions and the answers*. Orange County, CAS: Author.

Crawford, J. (1992). *Hold your tongue: Bilingualism and the politics of "English Only"*. Reading, MA: Addison-Wesley.

Dahlberg, J. (1995). *Smuggling people to the U.S. is big business in Thailand*. The Los Angeles Times, p. A1.

Darder, A., & Torres, R. D. (1998). Latinos and society: Culture, politics, and class. In: A. Darder. & R. Torres (Eds), *The Latino Studies Reader: Culture, Economy and Society*. Oxford: Blackwell.

Davis, M. (1990). City of quartz: Excavating the future in Los Angeles. London, New York: Verso.

del Olmo, F. (1997, September 14). End border hysteria and move on. *The Los Angeles Times*, p. M5.

Fields, B. (1990, May/June). Slavery, race, and ideology in the United States of America. *New Left Review*, *181*, 95–118.

Freeman, R. B. (1999). *The new inequality: Creating solutions for poor America*. Boston: Beacon.

Garcia y Griego, M., & Martin, P. (in press). *Immigration and integration in the post-187 era*. Berkeley: University of California Press.

Goldberg, D. T. (1993). *Racist culture: Philosophy and the politics of meaning*. Cambridge: Blackwell.

Guillaumin, C. (1995). *Racism, sexism, power and ideology*. London: Routledge.

Hall, S. (1990). Ethnicity: Identity and difference. *Radical America*, *13*(4), 9–20.

Harvey, D. (1989). *The condition of postmodernity*. Oxford: Blackwell.

Harvey, D. (1996). *Justice, nature, & the geography of difference*. Cambridge: Blackwell.

Horne, G. (1995). *Fire this time: The Watts uprising and the 1960s*. Charlottesville and London: University of Virginia Press.

Keil, R. (1998). *Los Angeles: Globalization, urbanization and social struggles*. New York: John Wiley & Sons.

Kyser, J. (1999). *Manufacturing in Los Angeles*. Los Angeles: Los Angeles County Economic Development Corporation.

Lee, D. (1999, July 26). L.A. County jobs surge since '93, but not wages. *The Los Angeles Times*, p. A1.

McLaren, P., & Torres, R. D. (1999). Racism and multicultural education: Rethinking "race" and "whiteness" in late capitalism. In: S. May (Ed.), *Critical Multiculturalism: Rethinking Multicultural and Antiracist Education*. London: Falmer.

Miles, R. (1989). *Racism*. London: Routledge.

Miles, R. (1993). *Racism after race relations*. London: Routledge.

Mirón, L. F. (1997). *Resisting discrimination: Affirmative strategies for principals and teachers*. Newbury Park, CA: Corwin.

Mirón, L. F., Inda, J. X., & Aguirre, J. K. (1998). Transnational migrants, cultural citizenship, and the politics of language in California. *Educational Policy, 12*(6), 659–681.

Muller, T. (1997). Nativism in the mid-1990s: Why now? In: J. F. Perea (Ed.), *Immigrants Out! The New Nativism an the Anti-immigrant Impulse in the United States*. New York: New York University Press.

The numbers game. (1994). In: The new face of America. *Time* [Special issue], *142*(21), 14–15.

Rifkin, J. (1995). *The end of work: The decline of the global labor force and the dawn of the post-market era*. New York: Tarcher/Putnam.

Sanchez, G. J. (1997). Face the nation: Race, immigration, and the rise of nativism in late twentieth century America. *International Migration Review, 31*.

Sassen, S. (1989). America's immigration "problem". *World Policy Journal, 6*, 811–832.

Sassen, S. (1991). *The global city*. Princeton, NJ: Princeton University Press.

Sassen, S. (1996, October). New employment regimes in cities: The impact on immigrant workers. *New Community, 22*(4), 579–594.

Scott, A. J., & Paul, A. S. (1991). Industrial development in Southern California, 1970–1987. In: J. F. Hart (Ed.), *Our Changing Cities*. Baltimore: The Johns Hopkins University Press.

Small, S. (1999). The contours of racialization: Structures, representations, and resistance in the United States. In: R. D. Torres, L. F. Mirón & J. X. Inda (Eds), *Race, Identity, and Citizenship: A Reader*. Oxford: Blackwell.

Soja, E. W. (1996). *Thirdspace: Journeys to Los Angeles and other real and imagined places*. Oxford: Blackwell.

Taylor, B. J. (1998). English for the children: A project of one nation/one California. Retrieved July 22, 1998 from the World Wide Web: http://www.onenation.org/aboutonoc.html

Torres, R. D., & Ngin, C. S. (1995). Racialized boundaries, class relations, and cultural politics: The Asian American and Latino experience. In: A. Darder (Ed.), *Culture and Difference: Critical Perspectives on the Bicultural Experience in the United States*. Westport: Bergin & Garvey.

Unz, R. K., & Tuchman, G. M. (1998). The Unz initiative. Retrieved July 22, 1998 from the World Wide Web: http://www.catesol.org/unztext.html

Valle, V., & Torres, R. D. (1993). The economic landscape of the Greater Eastside: Latino politics in "post-industrial" Los Angeles. In: *Prism*. Long Beach: Community Affairs Center, California State University, Long Beach.

Valle, V., & Torres, R. D. (1994). Latinos in a "post-industrial" disorder: Politics in a changing city. *Socialist Review, 23*(4), 1–28.

Valle, V., & Torres, R. D. (1995). The idea of Mestizaje and the "race" problematic: Racialized media discourse in a post-Fordist landscape. In: A. Darder (Ed.), *Culture and Difference: Critical Perspectives on the Bicultural Experience in the United States*. Westport: Bergin & Garvey.

Van Kemper, R., & Marcuse, P. (1997, November/December). A new spatial order in cities. *American Behavioral Scientist, 41*(3), 285–298.

Westphal, D. (1998, April 9). *Foreign-born population of U.S. surges*. The Orange County (Calif.) Register.

White, G. (1995, August 3). Workers held in near-slavery officials say. *The Los Angeles Times*, pp. A, A20.

Wieviorka, M. (1995). *The arena of racism*. London: Sage.

Wood, E. M. (1995). *Democracy against capitalism: Renewing historical materialism*. Cambridge: Cambridge University Press.

Wood, E. M. (1996, July/August). Modernity, postmodernity or capitalism? *Monthly Review, 48*(3).

Zukin, S. (1991). *Landscapes of power: From Detroit to Disney World*. Berkeley: University of Califronia Press.

THE ROLE OF EDUCATION IN COMMUNITY DEVELOPMENT: THE AKRON ENTERPRISE COMMUNITY INITIATIVE

Charis L. McGaughy

INTRODUCTION

The sustained interest in educational reform over the past two decades has focused attention on the complexity of problems facing distressed communities. Schools cannot improve the lives and performance of children in isolation. One implication of this realization is the sparked interest in community development from the educational arena. The re-emergence of community development within the social policy arena has created unique opportunities and challenges for educators to begin working with a wider array of individuals interested in improving the lives of children, families and communities.

This study examines the experiences and perceptions of educators engaged in a community development initiative in Akron, Ohio. Most significantly, the values of individuals involved with the educational programs will be examined. As will be described below, community development is underpinned by two major theoretical perspectives: Economic/Investment and Holistic Development. The merging of these two theoretical perspectives under one policy initiative raises difficult issues. Crowson, Wong & Aypay (2000) capture these difficulties with the following illuminating questions:

Community Development and School Reform, Volume 5, pages 121–138.
ISBN: 0-7623-0779-X

To what extent should or would community-level economic development (with its motives
of profit, property, employment, capitalization, and growth) drive the social? To what extent
might or can social capital formation (with its concerns for such elements as human welfare,
individual health, families, individual opportunity, lifelong learning, and civility) be incorporated effectively into the economic development of a community? (pp. 10–11).

The remainder of this paper examines how educators address these value
conflicts through a case study of a city actively engaged in a community development initiative. How did education gain such a significant role in the
community development initiative? What are the values, goals and experiences
of educators involved? What insights can be learned from their experiences?

The remainder of this paper explores these research questions. First, a brief
overview of the concept of community development is offered. This is followed
by a discussion of how the education reform movement became interested in
the concept of community development. Then, the dual theoretical underpinnings of community development, the Economic/Investment perspective and the
Holistic Development perspective, are defined. Next, the methodology and site
selection process used for this study are briefly detailed. Then, the federal
Empowerment Zone/Enterprise Community (EZ/EC) program and the Akron
Enterprise Community (EC) are discussed as specific examples of community
development. Finally, the findings and conclusions section of the paper explore
the context and perceptions related to education's role in the Akron EC.

OVERVIEW OF COMMUNITY DEVELOPMENT

The concept and goals of community development are wide-ranging, and the
process is malleable. After a lengthy analysis of the various definitions of
community development, Christenson et al. (1989) derived this succinct description: "a group of people in a locality initiating a social action process (i.e.
planned intervention) to change their economic, social, cultural, and/or environmental situation" (p. 14). The concept of community development is not
new. Beginning during the Progressive Era at the start of the twentieth century,
community development has fluctuated in popularity, and has reappeared in
several different policy incarnations. The Settlement Houses during the early
1900s, the New Deal during the 1930s, and the War on Poverty of the 1960s
are all examples of initiatives incorporating community development concepts
(for a detailed history, see Halpern, 1995). This broad-based concept over the
past century has been called on to "reduce class conflict, counter feelings of
anger and alienation, localize control of social institutions, create jobs and
reverse neighborhood decline, and address a variety of specific poverty-related

problems ranging from infant mortality to juvenile delinquency to overcrowded housing" (Halpern, 1995, p. 1).

EDUCATION'S INTEREST IN COMMUNITY DEVELOPMENT

The current interest in community development from the educational arena can be traced back to the report, "A Nation At Risk", issued by the National Commission on Excellence in Education in 1983. This report sparked a new era of school restructuring initiatives aimed at improving the American educational system, with the overlying goal of improving America's economic status. In response, three waves of educational reform have occurred over the last two decades, termed by Murphy and Adams (in press) as the Intensification era, the Restructuring era, and the Reformation era. The first wave, the Intensification era, was characterized by the push for top-down initiatives demanding stiffer standards for students and teachers that led to increased centralization. The next wave, the Restructuring era, was distinguished by the efforts to decentralize the educational system by using such methods as professional empowerment and site-based management. The third and current wave, the Reformation era, is distinguished by the movement to redesign the entire public school system by such means as increasing parental rights and privatization.

A major theme within the third wave of reform is the recognition that education alone cannot solve America's social and economic problems. As Hobbs (1995) explains:

> The failure of reforms and increased funding to be matched by improved performance led to a search beyond the school for reasons and they have been found. It has been found that changes in families and communities and the distribution of income cannot be separated from students' performance in school (p. 80).

The expectations placed upon schools have become overwhelming. Schools have become "default agencies" for addressing much broader societal issues than just academic performance (Timpane, 1997, p. 464). "Around the country," Dryfoos (1994) explains, "school administrators are crying for help. They acknowledge that they cannot attend to all the needs of the current crop of students and at the same time respond to the demands for quality education" (p. 4). This focus has led to the creation of an expanding base of literature "calling for an altered conception of 'the problem' of school improvement as a problem of community-wide change, not just school-based revitalization" (see Maeroff, 1998; Driscoll, 1995; Cohen, 1995a; Cibulka, 1996; Smrekar & Mawhinney, 1999; Smrekar, 2000, p. 3). School-community relations are receiving heightened attention and are being reconfigured in areas such as parental involvement, instructional partnerships, school-to-community "outreach", and children's services coordination (Crowson & Boyd, 1993, p. 140. See also Epstein, 1992; Henderson & Berla, 1997; Dryfoos, 1994; Cibulka & Kritek, 1996). In summary, educators are looking beyond the traditionally isolated institution of public education and seeking assistance and partnerships with other individuals interested in children and youth.

DUAL THEORETICAL UNDERPINNINGS

Two different theoretical perspectives underpin the current community development movement (see McGaughy, 2000). The first view, Economic/Investment theory, emphasizes using market forces to revitalize communities. The second view, the Holistic Development perspective, believes the entire ecology of the community needs to be supported in order for individuals to develop to their full potential. These two theories make dramatically different assumptions about the necessity and process of community development and education. The following descriptions delineate the two schools of thought:

Economic/Investment Perspective

This perspective originated with human capital theory developed by Schultz (1961). The argument is "people can be viewed as an economic asset in which increased investment in health, skills, and knowledge provide future returns to the economy through increases in labor productivity" (NEA, 1995, p. 13). Quality education is important because it is believed to increase labor productivity and foster economic growth. The 1983 report "A Nation At-Risk" championed this belief of the economic purpose for education. As Guthrie (1990) points out:

> 'Modern economics' is transforming schooling. Nations are attempting to enhance their economic position through the development of 'human capital', and, therefore, policymakers increasingly escalate their expectations for the performance of education systems (p. 110).

Beyond improving productivity, developing a high quality workforce in the Information Age is viewed as an economic imperative for remaining competitive:

> The future now belongs to societies that organize themselves for learning. What we know and can do holds the key to economic progress, just as command of natural resources once

did . . . More than ever before, nations that want high incomes and full employment must develop policies that emphasize the acquisition of knowledge and skills by everyone, not just a select few (Marshall & Tucker, 1992, p. xiii).

Education is viewed under this perspective as an investment in economic security. Public schools have received intense interest from the business arena, for "Only the schools can do the bulk of the job of educating young people" (Timpane, 1984, p. 390). Including education in community development, then, is seen as an important aspect of improving human capital and economic competitiveness. In addition, other sectors that develop human capital are also deemed important (such as housing, job training, child care and health services). Thus, the traditional focus of economic development is expanded to incorporate the needs of fostering a productive workforce.

Holistic Development Perspective

The goal of the Holistic Development perspective is broader than that of economic growth; the goal is to ensure that children, families and communities develop to their full potential. This view is "driven by a growing realization that many children face a complex set of overlapping and interrelated problems; changes in family patterns, demographics, and economic realities have created discontinuities between the needs of students and the abilities of both schools and families to meet them" (Smrekar, 1996a, p. 3). In order to meet these complex needs, communities need to be thought of as "ecosystems" (see Goodlad, 1987). Timpane (1997) succinctly describes the view as: "Borrowing language from the natural sciences, we believe that the arrangements we need are best thought of as an ecosystem – a total environment supporting the healthy growth and development of America's youth" (p. 465). Community development initiatives must take a holistic and inclusive approach, because in order for an initiative to succeed:

In the final analysis, no one institution is responsible for the collapse of the ecosystem for youth. And two corollaries come with this realization: (1) the reform or revitalization of any one institution cannot itself repair the ecosystem, and (2) no single institution can itself be reformed or revitalized without the concomitant strengthening of the other constituent parts of the system (Timpane, 1997, pp. 465–466).

This view also recognizes the importance of a healthy economy to a community. In Timpane's (1997) words: "Finally, were community development to accomplish all its goals for the healthy development of youth, its success would be for naught if young people were unable to find work in adulthood. Any community development must be linked to economic development" (p. 467).

METHODOLOGY AND SITE SELECTION

This study is a small piece of a much larger study (see Smrekar, 2000). Since the larger study is still in its infancy, this work offers preliminary findings from the first round of data collection in one of the larger study's three sites. This initial study uses qualitative methods of key informant interviews, observation, and artifact collection for data (see Strauss & Corbin, 1990; and Erlandson et al., 1993). We conducted fifteen semi-structured interviews with key informants, many more individuals were informally interviewed during the course of our observations, and all documents and statistics available related to the initiative were collected. We toured the participating community and visited six schools, a job training center, and school district and city offices. The intent was to develop a rich and descriptive analysis of individuals and activities involved with public education related activities within the community development initiative. This focus serves dual purposes: (1) to verify the validity of including this site in the larger sample; and (2) to obtain a significant data set of education related views, practices, and experiences in conjunction with community development.

Due to the preliminary nature of this report, there are several methodological limitations worthy of mention. Most significantly, this data set only represents one segment of the stakeholders that need to be interviewed. This weakens the credibility, transferability and dependability of the data (see Erlandson et al., 1993). In subsequent site visits, other stakeholders, such as community members (particularly parents), business people, social service providers and other individuals involved with the community development initiative will be interviewed. In addition, a quantitative component consisting of a survey will be conducted. In order to minimize the limitations of this data set, however, every effort has been made to triangulate any findings from the multiple sources available, and to limit the scope of the research questions to those applicable to the individuals interviewed.

The site selected for this study is a small Mid-Western city of about 250,000 people. In order to protect confidentiality no informant's name will be used. The first step used in identifying this site was the federal designation of an Empowerment Zone or Enterprise Community (EZ/EC) (see below for description) to guarantee an extant community development initiative. The original 105 Round I sites selected in 1994 were examined to ensure at least five years of community development activities were available. Including only those EZ/ECs with documented and confirmed participation by local K-12 public schools in the community development initiative then narrowed these sites down. Using the criteria of including a large urban site, a mid-size urban site, and a rural site for the larger study, the Akron EC was identified as the most promising mid-size urban site to study. This site was selected because of the unique approaches used in including public education in the community development initiative.

EMPOWERMENT ZONES/ENTERPRISE COMMUNITIES (EZ/EC)

In order to better understand the context of the Akron EC's community development initiative, a brief overview of the federal EZ/EC program is necessary. The EZ/EC is a federal initiative that currently supports community development. The EZ/EC idea originated in England, and was popularized in the United States by conservatives (i.e. Kemp's Urban Enterprise Zone legislation that failed during the 1980s). The belief was that economic renewal could be promoted in poor areas by using market strategies such as reducing taxes, employee expenses, and health and safety regulations (Stoesz, 1996, pp. 501–502). During the 1990s, President Clinton expanded upon this concept by adding a social services component. The result was the creation of the EZ/EC program in 1994. The mission is:

> To create self sustaining, long-term economic development in areas of pervasive poverty, unemployment, and general distress, and to demonstrate how distressed communities can achieve self-sufficiency through innovative and comprehensive strategic plans developed and implemented by alliances among private, public, and nonprofit entities (HUD, 1996a, p. 1).

During Round I of the program, six urban and three rural sites have been designated as Empowerment Zones. Over a ten-year period, the award offers $100 million each in social services block grants to the urban zones and $40 million each to the rural zones, plus a package of tax benefits, housing aid, and business incentives. Another 61 urban areas and 30 rural areas were chosen as Enterprise Communities and received $3 million each. Two other cities were awarded $215 million in economic-development grants for "urban supplemental zones" and four cities won slots as "enhanced enterprise communities" and received $25 million each (Cohen, 1995b, p. 1). In 1999, during Round II, an additional 15 urban sites and 5 rural sites have been selected as new Empowerment Zones, and 15 additional rural sites have been designated Enterprise Communities.

The framework of the EZ/EC program embodies four key principles:

- Economic Opportunity, including job creation within the community and throughout the region, entrepreneurial initiatives, small business expansion, and training for jobs that offer upward mobility;

- Sustainable Community Development, to advance the creation of livable and vibrant communities through comprehensive approaches that coordinate economic, physical, environmental, community, and human development;
- Community-Based Partnerships, involving participation of all segments of the community, including the political and governmental leadership, community groups, health and social service groups, environmental groups, religious organizations, the private and non-profit sectors, centers of learning, other community institutions, and individual citizens and;
- Strategic Vision for Change, which identifies what the community will become and a strategic map for revitalization (HUD, 1996b, p. 1).

The U.S. Department of Housing and Urban Development (HUD) reports that "What sets this Initiative apart from previous urban revitalization efforts is that community drives the decision-making" (HUD, 1996b, p. 1). The local governments have to develop and submit plans "showing how public and private institutions would work together on economic development, human services, transportation, housing, public safety, drug abuse, education and other concerns" (Cohen, 1994, p. 1). At this time, HUD reports that these EZ/ECs have generated over $4 billion in private and public investment to revive inner city neighborhoods, created approximately 20,000 jobs, and provided education and job training to nearly 45,000 people (HUD, 1999, p. 3).

OVERVIEW OF THE AKRON ENTERPRISE COMMUNITY

Located in the heart of the mid-West, Akron is a small city with a strong working class feel. The landscape is peppered with large industrial plants (some in operation, and others abandoned) and neighborhoods with older wood frame houses. This is also a city under transition. The once leading industry of rubber/tire production is being replaced by plastics/polymers. New suburban growth of fast food, strip malls, and housing developments rim the city. The overall population of the city is around 250,000.

The Akron Enterprise Community includes the downtown area, and the high poverty areas on the north, south, and west perimeters of downtown. The EC has 49,920 residents and covers and area of 13.5 square miles. The residents are 57.5% Non-Latino Caucasian, 39.5% African-American and 3% other. In the EC, 34.8% of the families live below the poverty level, compared to 16.5% of the families living within the city. Unemployment in the EC is more than double the rate for the rest of the city, with the most current estimates putting the rate at approximately 18%. Under the federal EC initiative, the Akron EC received $3 million in Title XX Social Service Block Grant funding beginning in 1994. In addition, the EC utilizes a mixture of federal, state and local resources to fund multiple projects within the EC.

The vision for the Akron EC is broad based. In short, the EC will "support entrepreneurial initiative, the expanding ability of businesses, the efforts of residents to achieve the skills necessary for job opportunities, and the development of neighborhoods which are secure, economically healthy and supportive of family life" (HUD, 1994, p. 1). The EC is addressing two major categories of need: Economic Opportunity and Sustainable Development. Table 1 lists examples of programs operating under each category (adapted from HUD, 1997).

Overall, the Akron EC utilizes a comprehensive approach for community development. In response, the initiative has received positive national attention. Housing and Urban Development Secretary Andrew Cuomo praised, "Akron's Enterprise Community is one of the best in the nation and can serve as a model for successful job creation and economic development" (HUD, 1998, p. 1).

FINDINGS AND CONCLUSION

In total, the findings below are derived from more than 20 hours of taped interviews resulting in 37 pages of transcribed dialog and notes. Site visit observation notes and all related documents collected are also used. All quotations below

Table 1. Akron EC Programs.

Economic Opportunity	Sustainable Development
• *Revolving Loan Program*: up to $75,000 loans available to EC business projects. • *Industrial Incubator*: technical assistance available for business growth and expansion. • *Microloan Program*: up to $5,000 for business start-ups. • *Polymer Training Center*: training program for current and future polymer workers.	• *Public Safety*: Police, Health, and Metro Housing Authority running education and law enforcement activities aimed at reducing drug activity. • *Urban Landscape*: initiated Capital Improvements program and opened national museum. • *Health*: increased immunizations and prenatal care, and opened Community Health Resources primary care facility. • *Supporting Families*: mental illness crisis center and in-home day care providers. • *Education*: school services coordination, tech prep, and economic enterprise curriculum. • *Housing*: low income housing construction, neighborhood improvement projects, and land reutilization program.

are taken directly from the transcripts. No names, only occupations, are listed of those quoted in order to protect confidentiality. Strauss and Corbin's (1990) method of open coding was used to analyze the data. Open coding is a "process of breaking down, examining, comparing, conceptualizing, and categorizing data" (p. 61). Through open coding, properties, dimensions and patterns in the data are discerned. First, however, in order to foster greater understanding of the context of this initiative, an in-depth description of the education-related EC activities is provided.

Education's Role in the Akron EC

Education has been a critical part of the Akron EC from the beginning of the initiative in 1994. According to a school district official, the mayor called him initially to inquire about what the city should do with the school district. When asked why the city was interested in including public schools, a city official explained, "Education initiatives from the very beginning were considered to be a very important part of what we wanted to do because we also took seriously the EC mission which [sic] was to be sustainable development." The original idea for education was to purchase a giant van that would travel to various schools and "demonstrate manufacturing processes" to students. The school district officials interviewed stated that they attended several meetings with city officials, and convinced them that the school district had different needs than the van.

Out of those discussions, the EC initiative involved public education in two areas. First, in response to the local demand for skilled labor, the city has established cooperative educational initiatives with the local university, businesses, and the public school district. There are six major career preparation initiatives sponsored by the EC (descriptions taken from correspondence received from an Akron city official dated 3/18/1998):

(1) *Middle School Technical Career Preparation*: Enables all middle school students in the EC to be exposed to workplace technology and investigate career opportunities for a full year. Students attend 9 week classes in each of these four areas: career planning; business communications; and two sections of technology and manufacturing in which they have hands-on experiences with Computer Aided Design (CAD), mills, lathes and other manufacturing productions. In addition, there are fieldtrips, classroom speakers and job shadowing.

(2) *Tech Prep*: A high school course of study that begins in freshman year, and is designed to develop technological skills. Classroom work is integrated towards workplace needs as identified by business advisors.

(3) *Polymer Internship Program*: This program is offered through the University of Akron's Akron Polymer Training Center for high school students who are interested in pursuing careers in plastics or rubber manufacturing. Students are selected to participate in a five week paid internship during which they are engaged in hands-on learning experiences. Students work in a processing lab and learn how to manufacture plastics using injection molding equipment and mills.

(4) *Technology Education*: Providing equipment to upgrade the World of Technology coursework offered to EC high school students. This program, in part, offers a continuation of the middle school program to enable students to keep up their skill development.

(5) *Urban Enterprise Education*: Coursework and training module was developed to teach EC area teachers in economics and the workplace and to use the Internet in economic coursework. This trains teachers to better teach EC students about the current world of work and business economics.

(6) *Student Workforce Training*: Working with EC high school students to establish a training-manufacturing environment, and develop internship and mentoring programs in conjunction.

Second, a social service component involving the 9 elementary and 3 middle schools located within the EC has been developed. This program has four targeted components (quoted from information provided by an Akron School District official):

- Increase social service agency involvement in target schools.
- Expand the availability of organized afterschool activities.
- Make meaningful parenting programs/self-improvement opportunities available to parents.
- Increase the number of families served by community health and social services agencies.

The family services program funds family service specialists who help connect individual families to social service providers, and afterschool and parental involvement programs. The individuals involved with these two program areas serve as the pool of key informants for this data set, which is analyzed in the next section.

RESULTS

This study seeks to explore the values and perceptions of individuals involved with the education-related activities within the Akron EC initiative. As Crowson, Wong, and Aypay (2000) point out:

At issue, from each side of the social and economic dyad, can be widely differing assumptions about: (a) market forces vs. services delivered routes toward the development of social capital, (b) the 'entrepreneurial spirit' of a community against the 'caring spirit' of a community, (c) the private-sector renewed against a public sector reaffirmed, and (d) 'development' interests in a community against 'welfare' interests (p. 12).

With these possible conflicts in mind, the property to be examined using open coding (see above for definition) is the theoretical perspective of the informants. This property is then broken down into dimensions, or the two theoretical underpinnings already identified above, the Economic/Investment perspective and the Holistic Development perspective. These dimensions are then examined for patterns. Table 2 demonstrates the framework for analysis.

Overall, what emerged from the data was an amazing alignment of the participants' goals, the specific programs with which they were involved, and the theoretical perspectives underpinning each program area.

The pattern of responses for those working with the career preparation area followed the Economic/Investment perspective. These individuals were primarily interested in making sure these students were ready for employment. Table 3 lists the occupations and some the various goals that the respondents stated.

As the goals listed in Table 3 overwhelmingly reflect, these educators want students to become productive workers. What was interesting was the nearly unanimous belief of these educators in the economic purpose of education. Not one individual noted any other goal beyond career development – no citizenship, quality of life, human development or democratic value goals were mentioned. This finding demonstrates the strength of the Economic/Investment perspective even within the field of education.

On the other hand, the pattern of the responses for those working with the school services program reflected the Holistic Development perspective. Again, this finding was not surprising, but the clarity of the purpose of education was not as well defined for these individuals. Table 4 illustrates the broad goals of the respondents.

Table 2.   Analytical Framework.

Property	Dimensions	Patterns
Theoretical Perspective	Economic/Investment ↔ Holistic Development	• Career Preparation: goals and perceptions. • School Services: goals and perceptions.

Table 3.   Career Preparation Goals.

Occupation	Goals
Teachers (4 interviewed)	• "My goal is to have students have a better grasp of career education and a better grasp of their toolbox for finding a job." • "My goal is to help students plant the seeds for the future." • "My goal is to familiarize the students with the terms and processes of a computer." • "My goal is to not make the students proficient in these areas, just try to give them experience, whether positive or negative."
School District Officials/ Program Directors (5 interviewed)	• "I would like the entire school to move in the direction of all students developing career paths." • "Ultimately, every kid will be prepared to work." • "There is the expectation that every student will do something." • There are three goals to the program: (1) make sure that students are aware of employment opportunities and the skills necessary; (2) help students to identify what training path to pursue; and (3) expose students to careers by providing hands-on experiences so that students can make informed decisions about career choices. • "In general, the program wanted to introduce students to careers. And if it was clear that the student was not going to college, the goal was to get him or her a job."

The variety of goals offered covers a wide range of activities, including everything from improving the quality of life, building community, and raising academic achievement. This reveals a disadvantage of the Holistic Development perspective because its goals are all encompassing. The Economic/Investment perspective has a much narrower, more targeted purpose that facilitates easier success and understanding. This is an issue that faces community development initiatives. By expanding the purview beyond traditional economic development, the increased complexity creates a multiplicity of challenges.

One argument in the community development literature is that "community development theory does not guide community development practice. Rather, practice is derived from the problems that emerge from the activities that the development professionals face without reference to any particular paradigm" (Christenson & Robinson, 1989). This assertion does not prove true in Akron. Theoretical perspectives do seem to drive the program's activities. In fact, on

Table 4. School Services Goals.

Occupation	Goals

School Level Personnel (3 respondents)	• "My goal is to save one child. If I can save more than one, that's a bonus. Actually, I would like to save them all. I need to set my goal low, so that when I save the 100 I am okay. It makes me happier."
	• "I think to create a link between schools, home and community that will provide a safe and nurturing environment."
	• One person or group cannot cover everything that needs to be done in schools. I recognize these following goals:
	
	  • Quality of life for students and families
	  • Communication between schools and home
	  • Better programs at schools so that children can be better served through increased access
	  • More involvement with community and parents
	  • More afterschool/evening programs
	  • School to be more community focused

School District Officials (2 respondents)	• "The bottom line was that there are a lot of issues beyond academics that make it hard for schools to be successful academically . . . If schools want to do the job of teaching kids, they need to focus on teachers and they had better be networking too with the community to do the other jobs." • "to build community, to provide a support system for families, positive interventions . . . What we are trying to do in accordance with all of the research that talks about building assets and linking kids to loving, caring adults, that kind of thing . . . We know that academics is kind of the ultimate goal of all of this that we raise achievement levels."

the surface, despite the differing value orientations of individuals involved in the two major education related areas, this does not seem to be creating any difficulties in Akron. Thus, the broad-based appeal of community development seems able to successfully incorporate fundamentally different ideological perspectives.

There were references made, however, to potential problems caused by the dualistic views. Concerns were expressed that the business sector was not adequately engaged with the school services component. The career preparation informants repeatedly commended the business community for their active involvement in everything from job shadowing opportunities and material donations to curriculum development. On the social services side, one individual explained the lack of business interest as "this sort of community-oriented good works thing they see as perhaps not, in this broad scope of issues, is perhaps not that great of interest to them, as opposed to say something like Tech Prep where I think they can see the obvious connection and the obvious benefits to them and their involvement." This reveals a risk involved with the Economic/Investment perspective of relying on business self-interest for participation. As Boger and Wegner (1996) warn, "over-reliance on the goal of economic self-interest may be unwise, since business ultimately may fill its needs in other ways – by locating its operations abroad or by importing skilled manpower form other nations" (p. 477).

This exposes an even larger problem community development initiatives face. The deeper problems that distressed communities must address are often beyond their power. Halpern (1995) points out:

> In a society characterized by social and racial segregation, the people coming together to address poverty and its related problems become the poor and excluded coming together to address these problems. The history of neighborhood initiative reflects a persistent tendency to ask those with the fewest capital, institutional, and human resources to draw on those resources to better their lives; to ask those whose trust has been betrayed over and over – notably African Americans – to join a process requiring significant trust; and to ask the excluded to be responsible for finding a way to become included (p. 12).

In order to be successful, community development must be nestled within a "national movement for growth with equity" (Fainstein, 1990, p. 44). Taking this larger view of community development as a national issue, this raises questions about the federal EZ/EC program. Is the program, with its piecemeal design of including a little over 100 communities, designed for only the possibility limited success? Are other communities that did not receive EZ/EC designation suffering because of "zero-sum" resource allocation?

Although the answers to these questions are beyond the scope of this study, Akron's experience does offer valuable insight. The designers of the Akron EC realized the limitations of $3 million in combating long-standing, persistent poverty. They purposefully selected activities that, as a city official describes, "would carry on or that we identified initiatives that could alter the landscape and the culture of the area and that are going to continue on as well." For example, the career preparation programs received start-up and infrastructure grants from the EC, and now the curriculum and staffing patterns have been altered within the school district to assure continuation. The social services component expects to continue also, although they face more challenges. Their funding streams have not yet become part of the school district's annual budget, but they have had some success in securing other outside grants. Therefore, in as much as the EZ/EC initiatives are able to establish long-term, systemic programming, the federal program can expect successes.

In the long run, the question becomes whether or not community development can sustain the interest and support of such disparate perspectives. In the words of Cibulka (1996), "Given this diversity of perspectives . . . what we have is a reform movement only in the broadest sense, one animated by a shared disdain for the status quo rather than a common ideological vision" (p. 404). The Akron EC offers insight in how to successfully resolve these differences. Not only have they managed to engage differing institutions within the community, such as education, social services and business, they have also seemed to ameliorate the gap between the theoretical perspectives. This has been accomplished by a strong unifying vision put forth by the initiative's leadership. This statement from a city official captures Akron's success:

> I don't think you can really have much community development until you at the same time providing [sic] good quality education that fits the needs of the persons being educated and fits the need of the community which needs those people as workers and as contributors to that community. And education not only in the sense of learning a skill but education in terms of development of the person, their character, their sense of responsibility, their ethics and then their competency on a broader range of skills. If you don't have that then your neighborhoods are not going to be good neighborhoods and your churches will not have people willing to be as involved and political as they are, and their voluntary organizations will not get as much support as they do, and certainly their businesses will suffer.

The main lesson learned from the Akron EC is one of hope. The hope is derived from the strong prospect of sustainability, found both programmatically and through leadership. The lesson seems so simple, yet in practice so difficult: keep the focus on meeting the affirmative, goal-oriented needs of the community.

REFERENCES

Boger, J. C., & Wegner, J. E. (Eds) (1996). *Race, Poverty and American Cities*. Chapel Hill, NC: The University of North Carolina Press.

Bourdieu, P. (1973). Cultural Reproduction and Social Reproduction. In: R. Brown (Ed.), *Knowledge, Education, and Cultural Change* (pp. 71–112). London: Tavistock.

Christenson, J. A., Fendley, K., & Robinson, J. W. (1989). Community Development. In: J. A. Christenson & J. W. Robinson (Eds), *Community Development in Perspective* (pp. 3–25). Ames, IA: Iowa State University Press.

Christenson, J. A., & Robinson, J. W. (Eds) (1989). *Community Development in Perspective*. Ames, IA: Iowa State University Press.

Cibulka, J. G. (1996). Conclusion: Toward an interpretation of school, family, and community connections: policy challenges. In: J. G. Cibulka & W. J. Kritek (Eds), *Coordination Among Schools, Families, and Communities: Prospects for Educational Reform* (pp. 403–435). Albany, NY: State University of New York Press.

Cibulka, J., & Kritek, W (Eds) (1996), *Coordination among Schools, Families, and Communities: Prospects for Educational Reform*. Albany, NY: State University of New York Press.

Cohen, D. L. (1994, September 21). Enterprise-Zone grants called boon to schools. Education Week on the Web [Online]. Available: http://www.edweek.org /edsearch.cfm/

Cohen, D. (1995a, May 3). Joining Hands. *Education Week*, 35–38.

Cohen, D. L. (1995b, January 11). Empowerment-Zone proposals prominently feature education. Education Week on the Web [Online]. Available: http://www.edweek.org /edsearch.cfm/

Coleman, J. (1988). Social capital in the creation of human capital. *American Journal of Sociology*, *94*(Supplement), 95–121.

Crowson, R. L., & Boyd, W. L. (1993, February). Coordinated services for children: Designing arks for storms and seas unknown. *American Journal of Education*, *101*, 140–179.

Crowson, R. L., Wong, K. K., & Aypay, A. (2000, May). The "quiet" reform in American education: Policy issues and opportunities in the school-to-work transition. *Educational Policy*, *14*(2), 241–258.

Driscoll, Mary (1995). Thinking like a fish: The implications of the image of school community for connections between parents and schools. In: P. W. Cookson, Jr. & B. Schneider (Eds), *Transforming Schools* (pp. 209–236). New York: Garland Publishing, Inc.

Driscoll, M., & Kerchner, C. T. (1999). The implications of social capital for schools, communities, and cities: Educational administration as if a sense of place mattered. In: J. Murphy & K. S. Louis (Eds), *Handbook of Research of Educational Administration* (2nd ed.). San Francisco: Jossey-Bass.

Dryfoos, J. G. (1984). *Full-Service Schools*. San Francisco: Jossey-Bass.

Epstein, Joyce L. (1992). School and family partnerships. In: M. Alkin (Ed.), *Encyclopedia of Educational Research* (6th ed.) (pp. 1139–1151). New York: MacMillan.

Erlandson, D., Harris, E., Skipper, B., & Allen, S. (1993). *Doing Naturalistic Inquiry. A Guide to Methods*. Newbury Park, CA: Sage.

Fainstein, S. S. (1990). The changing world economy and urban restructuring. In: D. Judd & M. Parkinson (Eds), *Leadership and Urban Regeneration: Cities in North America and Europe*. Newbury Park, CA: Sage Publications.

Fritch, W. (1999). Organizational Processes and Structures that Create and Maintain Social Capital in Selected Private and Public Schools. Doctoral dissertation, University of California Riverside.

Goodlad, J. I. (Ed.) (1987). *The Ecology of School Renewal: Eighty-sixth Yearbook of the National Society for the Study of Education*. Chicago: The University of Chicago Press

Guthrie, J. W. (1990). The evolving political economy of education and the implications for educational evaluation. *Educational Review*, *42*(2), 109–131.

Halpern, R. (1995). *Rebuilding the Inner City: A History of Neighborhood Initiatives to Address Poverty in the United States*. New York: Columbia University Press.

Henderson, A. T. & Berla, N. (Eds) (1997). *A New Generation of Evidence: The Family is Critical to Student Achievement*. Washington, DC: Center for Law and Education.

Hobbs, D. (1995, January). Human capital: The issues, enablers, and blocks in institutional change. Increasing Understanding of Public Problems and Policies. National Public Policy Education Conference (ERIC Document Reproduction Service No. ED386337).

Maeroff, Gene (1998, February). Altered destinies: Making life better for schoolchildren in need. *Phi Delta Kappan*, *79*(6), 425–432.

Marshall, R., & Tucker, M. (1992). *Thinking for a Living: Education and the Wealth of Nations*. New York: BasicBooks.

McGaughy, C. L. (Dec. 2000). Community development and education: A tripod approach to improving America. *The Urban Review*, *32*(4), 385–409.

Murphy, J., & Adams, J. (in press). Reforming America's Schools: 1980–2000. *Journal of Educational Administration*.

National Education Association (1995). How Education Spending Matters to Economic Development. West Haven, CT: NEA Professional Library (ERIC Document Reproduction Service No. ED389062).

Schultz, T. (1961). Investment in human capital. *American Economic Review, 51,* 1–17.

Smrekar, C. (1996a). The Kentucky family resource centers: The challenges of remaking family-school interactions. In: J. G. Cibulka & W. J. Kritek (Eds), *Coordination Among Schools, Families, and Communities: Prospects for Educational Reform* (pp. 3–25). Albany, NY: State University of New York Press.

Smrekar, C. (1996b). *The Impact of School Choice and Community.* Albany, NY: State University of New York Press.

Smrekar, C., & Mawhinney, H. (1999). Integrated services: Challenges in linking schools, families, and communities. In: J. Murphy & K. S. Louis (Eds), *Handbook of Research of Educational Administration* (2nd ed.) (pp. 443–461). San Francisco: Jossey-Bass.

Smrekar, C. (2000). Community development and the formation of policy environments for effective schooling. Grant application submitted to the Field-Initiated Studies Education Research Grant Program at the U.S. Department of Education. Nashville, TN: Vanderbilt University.

Stoesz, D. (1996). Poor Policy: The legacy of the Kerner Commission for Social Welfare. In: J. C. Boger & J. W. Wegner (Eds), *Race, Poverty and American Cities* (pp. 490–514). Chapel Hill, NC: The University of North Carolina Press.

Strauss, A. & Corbin, J. (1990). *Basics of Qualitative Research: Grounded Theory Procedures and Techniques.* Newbury Park, CA: Sage Press.

Timpane, M. (1984, February). Business has rediscovered the public schools. *Phi Delta Kappan, 65*(6), 389–392.

Timpane, M., & Reich, R. (1997, February). Revitalizing the ecosystem for youth. *Phi Delta Kappan, 78*(6), 464–470.

United States Department of Housing and Urban Development (1994). Strategic Plan Summary [Online]. Available: http://www.hud.gov/cpd/ezec/oh/akoh1.html

United States Department of Housing and Urban Development (1996a, March 16). About the initiative: Empowerment Zones and Enterprise Communities [Online]. Available: http://www.ezec.gov/About/ezfac.html

United States Department of Housing and Urban Development (1996b, March 16). EZ/EC fact sheet: Empowerment Zone/Enterprise Community Initiative [Online]. Available: http://www.hud.gov/cpd/ezec/ezecfct.html

United States Department of Housing and Urban Development (1997). Enterprise Community Performance Report 1995–1996 [Online]. Available: http:// www. hud.gov/cpd/ezec/oh/akoh-perf.html

United States Department of Housing and Urban Development (1999). Press Release: Cuomo delivers $1.48 million to Akron Enterprise Community and praises job creation and economic development efforts [Online]. Available: http:// www.hud.gov:80/pressrel/pr98-387.html

United States Department of Housing and Urban Development (1999). FY 2000 Budget Summary: Community Planning and development Empowerment Zones [Online]. Available: http://www.hud.gov/bdfy2000/ summary/cpd/ezec.html

LINKING COMMUNITY ORGANIZING AND SCHOOL REFORM: A COMPARATIVE ANALYSIS

Dennis Shirley

In the field of school reform, the limitations of interprofessional collaboratives to remove the barriers to children's learning have become more evident with the passage of time (Schorr, 1997). As the efficacy of interprofessional collaboratives has been called into question, a more expansive conceptualization of collaboration has been brought to bear on the problems of urban schools and communities. Community-based organizations with a variety of linkages to voluntary associations led by "non-professional" or working class indigenous populations have begun to explore the role that they can play in the improvement of curricula, governance, and assessment in public schools (Shirley, 1997).

These new departures in community organizing and school reform have dovetailed with recent advances in the social sciences that focus on the elaboration of *social capital* theory. In contrast to *financial capital*, which designates the value of money, or *human capital*, which refers to an individual's intellectual and physical capabilities, *social capital* describes the kinds of capital that are embedded in social relationships. A grandmother or an aunt who provides child care in the afternoon for children while their parents work reflects one *informal* kind of social capital, while social capital can also be *formalized* through institutional arrangements such as the Girl Scouts, a PTA, or a congregation.

In addition, social scientists also distinguish between *bonding social capital* – which occurs among individuals in an institutional setting such as a church or trade

Community Development and School Reform, Volume 5, pages 139–169.
Copyright © 2001 by Elsevier Science Ltd.
All rights of reproduction in any form reserved.
ISBN: 0-7623-0779-X

union – and *bridging social capital*, which refers to strong lateral ties between individuals that cross organizational boundaries. In terms of schools, I will refer to bonding social capital as that which strengthens social ties between teachers or between teachers and students, whereas bridging social capital will designate ties that can be created between teachers, clergy, and community activists. The distinction is particularly instructive when one contrasts interprofessional collaboration with community organizing approaches, for the former emphasizes social ties between professionals (a bonding approach) whereas the latter stresses ties between indigenous community members and the professionals who control schools and affiliated service institutions (the bridging alternative).

Social capital theory has recently been brought to bear on a wide number of topics confronting American society, ranging from low voting patterns to school failure to crime. At first developed tentatively by economist Glenn Loury, sociologist James Coleman then developed the idea with more rigor, and applied it to explain what he perceived as the superior achievement of Catholic schools in educating low-income urban youth. It was not until political scientist Robert Putnam published his now famous essay, "Bowling Alone," however, that social capital theory gained widespread popularity as an explanation for fin-de-siècle malaise in the American body politic (Coleman, Hoffer & Kilgore, 1982; Coleman & Hoffer, 1987; Coleman, 1988, 1990; Loury, 1977, 1978; Putnam, 1993, 1995).

Putnam's "Bowling Alone" identified a widespread disengagement of Americans from civic institutions and social life. Putnam suggested that this disengagement both contributed to a drop in voting, participation in mediating institutions such as churches and civic clubs, as well as a decline in social trust. In its most dramatic expression, social decapitalization appeared to reflect the problems of a society that had lost its moral bearings and cultivated individualism and materialism at the expense of sociability and an ethic of mutual responsibility.

Putnam's "Bowling Alone" was easily the most influential monograph of its type to appear in the 1990s and spawned a wave of scholarly research on social capital. It is intriguing to note that even when critics attacked Putnam's findings they generally did not question the utility of social capital theory as a framework for evaluating social problems. Andrew Greeley, for example, took Putnam to task for failing to acknowledge the concentration of social capital in religious institutions in the United States, but explicitly sought to refine social capital theory in the process. Similarly, Alejandro Portes criticized Putnam's use of the concept for its "logical circularity," believing that Putnam construed the concept in such a way that it was both a cause and a result of virtually every social problem in contemporary American society. Like Greeley, Portes sought to circumscribe the boundaries of social capital more clearly to ensure that the notion would maintain its conceptual vigor (Greeley, 1997; Portes, 1998).

This chapter seeks to maintain this spirit of creative reconceptualization and development of social capital theory, in this case in the instance of education and community engagement. Recognizing that social capital theory is in the infancy of its development – just as human capital theory was a generation ago – we must anticipate that only continued theoretical and empirical work will clarify whether social capital will develop into a truly robust concept that can clarify hitherto obscure facets of social life. In this chapter I endeavor to use social capital theory to explore contemporary efforts in the overlapping terrain of school reform and community organizing in two South Texas schools and communities. I also seek to criticize and refine social capital theory by distinguishing between different kinds of school and community ties and underlining their varying ramifications. Some theoretical rigor in disaggregating robust social ties from more ephemeral relationships is of some urgency if we are to develop public schools into centers of civic engagement rather than islands of bureaucracy that are more devoted to district and state mandates than the communities they serve.

In the two case studies that follow, we shall explore whether the concept of social capital can bear the amount of theoretical weight attributed to it in recent contributions to the social sciences. If parents become more engaged in their children's school and in community development, does that engagement necessarily translate into increased academic achievement for children? James Coleman suggested that social capital explained the superior outcomes from Catholic school education over public schools in inner city communities, but subsequent commentators have noted that Coleman's analysis was flawed, for two reasons. First, Coleman's study focused on academic achievement at the high school level, but the youths in his study typically did not come from a parish in close proximity to a school, but rather from a scattered metropolitan area. Second, the parents were more or less involved in a fiducial relationship with the school; they trusted and supported the school's authority, but were not otherwise particularly engaged with it (Bryk, Lee & Holland, 1993).

In spite of this criticism of the fundamental premises of Coleman's work, the notion that social capitalization could cause high levels of academic achievement has continued to provoke high levels of interest among sociologists of education. In the two cases, however, a more complex portrait emerges – one in which social capital formation has many benefits but in which a pattern of causation in terms of raising academic achievement is ambiguous. These findings lead one to suggest that social capital formation has an intrinsic worth in terms of its contributions to human relationships and community development, but should not be viewed as a panacea for either the multitude of problems that confront underresourced communities in general or for issues relating to student academic achievement in particular.

We shall also explore what might be referred to as "hidden costs of social capitalization." Extensive research indicates that school reform efforts typically assign classroom teachers hosts of new responsibilities without any corresponding relief from the already demanding tasks of planning instruction, teaching classes, and correcting student work. This phenomenon of "intensification" can lead teachers to lose a focus on student learning, expend valuable time on issues peripheral to their vocation, and in extreme cases, leave their profession. If social capital theory is to make a lasting contribution to school reform, it must recognize the problems inherent in intensification and develop strategies to assure that the instruction of classroom teachers is not compromised, but enhanced (Sarason, 1990; Hargreaves, 1994; Muncey & McQuillan, 1996).

Although social capital theory is in need of further elaboration, even in its current incarnation it offers a crucial and often missing analytical tool for those who are concerned with the political and economic problems that characterize contemporary American civil society. Rather than the reigning methodological individualism, which studies institutions and social phenomena in relative isolation from one another, social capital theory is interested in the nexus between institutions, cultures, and group norms. This last point is particularly important, because by recognizing that values infuse and shape social relationships, social capital theory challenges both functionalist and structuralist schools of social analysis and restores the centrality of norms to social inquiry. Finally, social capital theory, with its explicit interest in secondary associations and civil society, provides a heuristic device that can diagnose problematic areas in which social networks have weakened and, at least potentially, lead to strategies which will strengthen those networks that exist in the interstices beyond the boundaries of both the state and capitalism. It can thus generate new forms of knowledge in the key intermediate zone of "civil society" which entails families, religious institutions, and neighborhood associations and exists between the bureaucracies of the welfare state and the vagaries of the marketplace (Arato & Choen, 1992; Hall, 1995).

The Alliance Schools

One of the most intriguing educational developments in the United States in the 1990s was the formation of a state-wide network of "Alliance Schools" affiliated with the community organizations of the Industrial Areas Foundation

(IAF) in Texas. The Industrial Areas Foundation was founded by Saul Alinsky in Chicago in 1940 and served as the training ground for many of the nation's preeminent social activists in the second half of the twentieth century. While "Alinsky organizations" maintain a reputation as radical groups bent on social change, the organizations evolved considerably in the last quarter century. Alinsky's emphasis on flamboyant radicalism was jettisoned after his death in 1972, and his successor, Ed Chambers, deliberately sought out political pragmatists to lead the group and focussed on basing the organizations in congregations (Shirley, 1997).

Alinsky organizations generally had their strongholds in African-American neighborhoods in northern cities such as Chicago, Rochester, and New York through their first thirty years, although fledgling organizations in Hispanic barrios in California also arose in this time. It was not until 1972, however, when systemic organizing efforts began in Texas for the first time. Launched by Ernie Cortés in 1972, a new group called Communities Organized for Public Service (COPS) brought hitherto disconnected barrios on the West and South Sides of San Antonio into the political arena. COPS fought hard and well against the well-established Good Government League in San Antonio, and won victories in the form of new libraries, parks, school crossing guards, drainage, paved roads, and a host of other infrastructural improvements in San Antonio's barrios.

The successes of COPS in San Antonio soon became well-known throughout the Southwest, and a host of sister organizations sprang up in Houston, El Paso, Austin, and the Rio Grande Valley. "Valley Interfaith" was the name of the IAF group that was started in the Lower Border area of Texas in 1983, and which began by attacking problems resulting from inadequate sewer lines, lack of access to clean drinking water, and unpaved roads in impoverished "colonias" or unincorporated rural settlements in hamlets that lined the northern flank of the Rio Grande. Valley Interfaith quickly became involved in state-wide school reform efforts, which culminated in a large legislative package known as "House Bill 72" which reallocated millions of dollars of funding to property-poor school districts in the Lone Star State.

By the early 1990s the Texas IAF had become a powerful political force in Texas. In terms of schools, the IAF had won a particularly impressive victory by turning around a crisis-ridden middle school in Fort Worth, and similar school improvement initatives were explored in Houston, El Paso, Austin, and San Antonio. The IAF then pushed Texas Commissioner of Education Lionel "Skip" Meno to create funding for a bold new network of urban schools affiliated with the IAF network. Called "Alliance Schools," these schools are based in the poorest communities in Texas and have grown from an initial cluster of 21 schools to 142 at the time of this writing (April, 2001).

How have Alliance Schools developed the civic capacity of parents, teachers, and community members to improve their public schools? In some schools, a glaring problem, such as rats in the library or drug dealers on the front steps of the school after hours, have served as the point of departure for an organizing effort. In other settings an innovative principal, or an enterprising and charismatic minister, have sparked a community's interest in improving a school and enhancing the quality of life in a neighborhood. There is no one way to launch an Alliance School. Yet there does seem to be a pattern insofar as all of the schools have abundant social problems which can stir the outrage of parents and teachers and lead to concerted efforts at confrontation with power holders and intense negotiations for improved conditions.

The case studies illustrate the trajectories of two schools which have joined the Alliance School network. The cases both indicate that school and community development can happen in very real terms in Alliance Schools, and they also demonstrate that this is very hard work indeed, with no guarantees and plenty of opportunities to undermine social capital formation. Yet the overall impression is one of positive change, even if much of it is contested and still very much evolving.

Sam Houston Elementary School

Our first case study concerns Sam Houston Elementary School in the barrio of "La Paloma" in south-central McAllen, Texas. Although McAllen is a small city by most measures, Sam Houston's immediate environment is in many ways similar to barrios in larger cities, such as the West Side of San Antonio or the East Side of Austin; tiny shacks packed with immigrant families alternated with more stable middle-class homes, and roosters and goats could be seen in small back yards or wandering into the streets, indicating just how close austensibly urban residents remained to the pace and texture of rural life. In the early 1990s, Sam Houston served a low-income Mexican American community that witnessed scant evidence of the growing prosperity of McAllen's North Side. First opened in 1920, Sam Houston was the oldest public school in McAllen. Regretfully, the district had allowed the building to run down, and it was badly in need of infrastructural repairs.

The principal of Sam Houston, Connie Maheshwari (born Connie Anaya of Mexican-descent parents, she married an Indian immigrant) entered teaching in 1979 at Wilson Elementary School and was the assistant principal at Rayburn Elementary School in McAllen from 1987 to 1990. She was appointed principal of Sam Houston in the fall of 1990 and spent several years developing her leadership skills and struggling to improve the academic achievements of the over five hundred children that attended Sam Houston. She first became interested in becoming an Alliance School in a partnership with Valley Interfaith when Father Bart Flaat of Saint Joseph's the Worker Catholic Church in south McAllen set up a meeting to promote the new school network in McAllen in January 1994.

Father Bart had been active with COPS in San Antonio from 1977 to 1985, where he was recruited into the community-based organization by Robert Rivera, the lead organizer at that time. When Father Bart accepted a position at Our Lady of the Assumption in Harlingen in 1985, he spent two years working with his parishioners to develop their trust in Valley Interfaith before they agreed to become dues-paying members. Yet even when they did join, the cautious congregation never developed the civic leadership of which Father Bart was persuaded they were capable.

Father Bart discovered a similar atmosphere at Saint Joseph's when he took over its leadership in 1991. Although Saint Joseph's supported Valley Interfaith, Father Bart considered it to be a "sleeping member" which rarely organized in any active way to support the agenda of the community-based organization. Yet Father Bart sensed a hunger for change among many individuals with whom he had his first contacts in his parish, which comprises a number of neighborhoods – La Paloma, Hermosa, Balboa, Alta Linda, and Los Encinos – in south McAllen. To ascertain the depth of such sentiments, Father Bart began a series of conversations in the community.

For Father Bart, the issues that emerged from those conversations were twofold, and resulted in a parish development program that he established with the most active church leaders. "First, we wanted a clinic for people who fell between the cracks – who had neither medicaid nor health insurance – and we learned that there were loads of people like this in south McAllen. Second, we heard a lot about schools and education. A lot of our parents have very little education, and schools are very intimidating institutions for them. And our parents wanted to get involved, but they wanted to do more just than folding papers at the school" (personal communication, May 21, 1997).

Saint Joseph's the Worker became a lively religious community in the months following Father Bart's arrival. He established thirty "comunidades de base" in his parish, where laity met on a weekly basis, discussed troubles in their neighborhood, and explored passages from scripture to explicate their relevance to their challenges. They discussed crime, unemployment, and family problems, and Father Bart worked with his leading parishioners to interpret their concerns in light of Catholic theology. "Whenever you see people who are oppressed because of an economic system, our scripture gives us a very clear message that we need to empower people, to help them makes sense of their lives, and to help them to gain the power to make decisions and to reach out in new kinds of ways," Father Bart commented.

The heightened engagement with Catholic theology in the parish of Saint Joseph's resulted in concrete forms of political action. Parishioners educated themselves on the importance of voting and supported voter registration drives organized by Valley Interfaith in south McAllen. They agitated for improved health care, and collaborated with city council to create a clinic called "El Milagro" to support the needs of low-income citizens and immigrants who otherwise could not access affordable medical care. Many parishioners were alarmed by increasingly brazen acts of crime in their neighborhood – which resulted in the deaths of two youths and a police officer in the summer of 1993 – and they began a series of meetings with the police to enhance security. Concerning the community's desire for improved education, Father Bart was intrigued to learn of the creation of the Alliance Schools in the summer of 1992, and for the following year he promoted the new network in his discussions with public school principals in south McAllen. Another round of roughly fifty house meetings was convened throughout La Paloma at the same time – using the comunidades de base as the forum for announcing and organizing the meetings – to discuss the emerging political agenda for the community.

All of these different organizing initiatives came together in a "Parish Convention" held at Saint Joseph's in September 1993, attended by over six hundred parishioners and community residents, where a host of different issues, engendered through months of individual conversations and house meetings, coalesced into a coherent program of neighborhood development to promote the well-being of the families living in the South Side. Parishioners had invited in several public officials, such as school board members, city commissioners, and the chief of police, and they received commitments from the officials to help them with problems such as unpaved roads, drive-by shootings, and disrepair of school buildings.

The next step was a meeting set up by Father Bart to convene all of the public school principals in McAllen to discuss in greater detail the possibilites and risks that could attend becoming an Alliance School. At the meeting, some principals expressed concern that collaboration with Valley Interfaith could be interpreted by families as mixing politics – and radical politics at that – with the public schools. Others were worried about collaborating with the churches that provide the dues-paying backbone of Valley Interfaith, and raised issues about the separation of church and state. For several, however, the possibilities seemed to outweigh the risks. Connie Maheshwari was one of the principals invited to the meeting; she had previous experience with Valley Interfaith around the issue of housing in the Valley, and the possible collaborative struck her as one that could benefit her pupils and their families.

When Maheshwari returned to her campus after the meeting and began discussing the idea of the Alliance Schools with her staff, she found that some teachers shared some of the concerns raised by principals at the meeting convened by Father Bart. Teachers were particularly concerned that the Alliance Schools collaborative could entail a loss of their sense of professional autonomy. "Teachers are afraid that Valley Interfaith is going to come in and tell them what to do," Maheshwari recalled. "This is a real fear." She also noted that several teachers were worried that empowered parents "could become a monster" who might at first conduct purposeful advocacy for their children but then develop an adversarial tone with teachers. It took her several months to work patiently with her staff to persuade them that the Alliance School concept represented a calculated risk which could benefit their students by improving relationships between their homes and the school (C. Maheshwari, personal communication, December 16, 1997).

Throughout the fall and winter of 1994 Valley Interfaith organizers, Sam Houston teachers, and indigenous community residents conducted scores of meetings in the neighborhood surrounding Sam Houston. Parents complained about poor lighting, lack of supervision in the numerous back alleys which students took to and from school, and abandoned homes where teenagers met to sell and use drugs in close proximity to Sam Houston. Many parents, and especially single mothers, were concerned that they had to work full time and had no way to supervise their children in the late afternoon. Other parents were simply worried about the abundance of trash – old tires, broken glass, rain-soaked mattresses – which littered the streets and alleys around the school and seemed to escape the attention of city sanitation workers. Finally, teachers from Sam Houston who attended the house meetings and those parents who were active in the school shared their concerns about the crumbling physical infrastructure of the school, the persistent presence of rats in classrooms and the cafeteria, and their hope for a new building.

"I loved those house meetings," Connie Maheshwari recalled, "because that's where I really learned about our community, and that's where our teachers began learning about the condition of children in our community. That's where we really started putting together a genuine curriculum unit that would enable us to become a true community of learners" (personal communication, December 16, 1997). Teachers were happy when parents who had been reluctant to come to the school approached them at house meetings, asked about ways that they could support their children's learning, and began coaching their children at home in ways that had real benefits for the children's academic achievement. Slowly the ties of bridging social capital between the teachers and the parents were built through the house meeting strategy.

One of the important developmental processes that happened through all of these actions is that parents and teachers were working together to improve educational conditions for Sam Houston's children, and they began to appreciate the talents that each group brought to the process. "I remember going with the parents to meet with the chief of police, Alex Longoria," Maheshwari said. "But I didn't lead the meeting. I simply listened to what the parents were saying. They did all of the talking, and they're the ones who had some real ideas for change worked out in advance. And once you've been through several meetings like that, you start to change and to understand that parents have strengths that somehow you never really saw or understood before" (personal communication, December 16, 1997).

All of the meetings began to crescendo in January 1995, when the community worked with Valley Interfaith to prepare a large public assembly – which they called a "Kids' Action Assembly" – to create a greater climate of community accountability for the children. Estela Sosa-Garza of Valley Interfaith worked closely with parents and teachers, role-playing the statements that they wanted to make and the questions they wished to address to public officials. According to IAF community organizing traditions, even if one has prior agreements from public officials to work together, it is crucial that those agreements take on a public texture before low-income communities. Those large assemblies mirror back to the community progress that has been made through many months of political organizing, and they also demonstrate the leadership that has been developed by indigenous community residents.

Parents, teachers, and Valley Interfaith organizers contacted numerous public officials to come to the assembly – scheduled for February 1995 – to commit to improved educational conditions. Public officials such as the chief of police, the director of Parks and Recreation, city commissioners, the city manager, the superintendent of schools and school board members were all informed in advance as to the nature of the assembly and the kinds of questions which would be posed to them. When the evening of the Kids' Action Assembly finally arrived, over three hundred parents from the school attended – a hitherto unprecedented gathering by the community on behalf of its children. Raquel Guzman – a teacher – introduced the assembly in English, and David Gomez – a parent – did the same in Spanish. The public officials heard parent leaders such as Delia Villarreal, Christina Fuentes and David Gomez, as well as teachers such as Leticia Casas, Raquel Guzman, and Mary Vela, who described the problems in the neighborhood and asked for their leadership in improving Sam Houston and its community. Speaking in a mixture of Spanish and English, the parents and teachers together committed themselves to working together and to demanding accountability from their civic leaders.

That Kids' Action Assembly played a pivotal role in the history of Sam Houston. For the first time, members of the community saw a host of leaders who were made up of their friends and neighbors seeking a new relationship with public officials and garnering results. As a consequence of the meeting, the city department of Parks and Recreation agreed to fund an after-school program for Sam Houston, which enrolled over two hundred children in its first year. A police substation was opened closer to the school and additional officers were assigned to patrol the area. City commissioners made sure that the trash which filled the alleys near the school were cleaned up and additional lighting was installed. And to make sure that the community developed its own capacity to improve its children's education, parents at Sam Houston signed a "parent contract" in which parents agreed to ask their children over dinner about their day in school and to insist that homework be done punctually. By May 1995 the school and its community seemed indeed to have turned over a new leaf, and the self-confidence of both the parents and the children seemed to have enhanced immeasurably. The only measure not substantially addressed concerned the community's hopes for extensive renovations at Sam Houston, if not a brand new building.

As can so often happen with the uncertain politics of school reform, however, circumstances developed in such a way that a surprise ending awaited the advocates of a new building. Just before the Kids' Action Assembly, the McAllen Independent School District hired a new superintendent, Dr. Robert Schumacher. Schumacher inherited a joint plan between the school district, the city of McAllen, and a private corporation called "McAllen Affordable Homes" to build a new neighborhood which would place home ownership within the grasp of moderate and low-income city residents. Called "Los Encinos," the neighborhood was planned several miles south of McAllen proper. It would consist of 246 new homes and the McAllen school district had committed to construct a brand new elementary school there, fully equipped with fibre optic cables to enable students to take full advantage of computer technology.

In January 1996 Schumacher proposed that part of the Sam Houston student body be bused to the new campus and that the remaining children be reassigned to Zavala and Tipton, two elementary schools close to La Paloma. Instantly the community was split about the advantages and disadvantages of the proposal. While many were enthusiastic about the capabilities of the new campus and the prospect that their children would benefit from what appeared to be a state-of-the-art building, others were disappointed at the school's remove from its community and the scattering of the student body to three different campuses. Some even suggested that the new school was a way to split a community when it had become empowered to advocate successfully for its children. Others, however, pointed out that many other parents in McAllen would be eager to have their children attend the new school and that to squander this opportunity would be a major disservice to the children. This latter group ultimately prevailed, and Sam Houston Elementary School was allocated a brand new building, surrounding with ample field for playgrounds and ball parks.

For many educators the most intense phase of struggle at Sam Houston could now appear more or less terminated. La Paloma had garnered a brand new, wonderfully equipped campus for Sam Houston. Yet part of the victory was phyrric, for only one-fifth of the students on the new campus came from the old Sam Houston; the other students elected to attend neighborhood schools closer to home than to venture six miles south to Los Encinos. Determined to make the most of a difficult situation, Connie Maheshwari struggled to keep up as high a level of parental engagement as possible while ensuring that teachers in the school did all within their power to develop a curriculum that was linked to the community and preparing the children to exercise the same kind of civic leadership that had been modelled by their parents. In the two years after Sam Houston's move, Maheshwari and her staff have implemented a number of curricular innovations which have sustained an unusually rich relationship between the school and its new, extended community.

One of the intriguing developments that occured in the school in its new setting was the implementation of a project-based approach to curriculum development. Several teachers at Sam Houston had received training from the McAllen Independent School District on the development of interdisciplinary curricula, and the teachers were eager to use their training to pilot units of study that would be relevant to the children's community. Yet since so many children at the school came from the brand new neighborhood, it would seem difficult to develop a course of study around a settlement with little of the cultural fabric that develops over time in more established neighborhoods.

The response of the teachers to this curricular dilemma was to engage the students in conversations about their community. Early elementary teachers such as Diana Garza and Raquel Guzman began asking their pupils if they observed anything noteworthy in Los Encinos that they might want to learn more about. The children instantly responded that there was a great deal of housing construction in the community. Observing high levels of student interest in this phenomenon, the teachers asked their pupils if they would like to learn more about how a home is built. When the children responded positively, the teachers began to develop a curriculum that would take the children into the heart of the new neighborhood and would give them a broad range of academic skills which they could apply to a multitude of disciplines.

First grade teacher Raquel Guzman decided to explore a project on construction as part of her regular classroom instruction. Once she had inquired whether the children would like to learn more about building homes, she worked with the school librarian to acquire books on construction, plumbing, and electricity, and she asked the foreman of construction across the street from the school if he would be willing to talk with the children about his work. When the foreman obliged, he discovered that the children had many questions that he could only explain with difficulty without visual reference to a new home. He then invited the class to go with him to the construction site, where he explained all of the planning and skill that must go into building a house.

Raquel Guzman was particularly intrigued to observe the way that the project on construction catalyzed the learning of children who previously were marginal in their academic achievement. She noted that one particular student usually needed extra attention when it came to reading, writing, or mathematics. Guzman admired his tenacious efforts to succeed academically, but she also observed that his struggles to master the regular curriculum were damaging his confidence. Once the project on home construction began, however, "He lit up. He was always the first one who knew all of the new vocabulary, while he usually wasn't so confident because he hadn't been a strong reader and always needed extra help. I was so surprised with him, and he's made much more of a connection to me. Best of all, the new skills he has learned have spread to the other academic areas and the other kids now see him as a real class leader" (personal communication, Raquel Guzman, 5 April 1997).

Guzman observed that the construction project has also helped the children to make stronger connections to their own community, even if they are bused to the new campus from the old Sam Houston neighborhood. Similar connections have been made by other enterprising teachers. In the fall of 1997 teachers Leticia Casas, Mary Vela, Susanna Sarmiento, and Mary Ann Rosales taught fourth and fifth graders about an upcoming school bond election, scheduled for 4 October. The teachers taught the students about the issues at stake in the election and prepared them to venture forth into the community to inform neighborhood residents about the referendum. As part of the preparation students met in small groups and wrote out exactly what they would say to community members in both Spanish and English. Following role-play procedures that teachers had learned from Valley Interfaith organizer Estela Sosa-Garza in preparation for the Kids' Action Assembly, students read from their scripts and entertained possible responses from the community so that they would be ready to answer questions in an informed and accurate way.

Another manner in which Sam Houston has changed its culture concerns the manner in which it has been resourceful about utilizing the talents of parents as curriculum developers. Continuing the tradition of house meetings begun at the time of the Kids' Action Assembly, teachers have met with parents on a regular basis in their homes to discuss innovative course materials which they could design together for the students. One immigrant mother, Eusebia Hinojosa, expended tremendous energy to prepare Sam Houston for a celebration of Mexican independence day – on 16 September – in the fall of 1997. Hinojosa rallied a group of other parents around her, sought out curriculum materials on the Mexican side of the border, and taught classes that demonstrated folk art and indigenous dance traditions to the children.

As a result of these developments, Sam Houston has developed a self-sustaining momentum in the school and community in which parents and teachers meet on a regular basis to develop culturally responsive pedagogies that will enable the children to hold on to key facets of traditional Mexican culture at the same time that they acquire the civic skills that will allow them to participate in the political arena in the United States. Mothers such as Eusebia Hinojosa, Irma Hernandez, and Monserrat Herrera have worked with teacher to develop curricular units celebrating el Día de los Muertos, el Día de los Reyes, and el Día de la Candelaria, which are all Mexican holidays that enfuse indigenous and Christian religious practices in Mexican culture.

Through these different efforts, Sam Houston is finding myriad pedagogical channels which can sustain and enliven the organizing efforts that began in 1994 and which reconstituted the school as a civic center for its community. Although the process of relocating the campus on the far southern edge of McAllen was contested, teachers, students, and parents have found ways to work together that have engendered new curricular units, developed the role of parents as pedagogical innovators, and taught students the practical nuts-and-bolts of civic engagement. In additional to all of this educational ferment, the teachers have also sought to be realistic and to link students' learning, when-ever possible, to the skills that are measured on the Texas Assessment of Academic Skills (TAAS) test, a criterion-referenced test administered throughout the Lone Star State. Thus, teachers helped students to summarize newspaper articles on the bond referendum in a manner similar to that which appears on the TAAS and students wrote about Mexican folk traditions in a fashion which helped them to monitor and reflect critically on their own learning. "We don't completely detach ourselves from TAAS objectives," Connie Maheshwari said. "In fact, we work very hard on the problem-solving skills that are needed for the test. We just do it differently" (personal commu-nication, December 16, 1997).

That gamble – the wager that one could both develop a full range of pedagogical and curricular innovations and do well on the TAAS – is one which few educators in Texas have been willing to take in recent years. The costs of doing poorly on the TAAS have simply been too high to allow educators much margin for error while exploring new approaches to teaching and learning. Yet Sam Houston was to experience a different and more felicitous outcome when it received its TAAS scores in the spring of 1998, and the faculty learned that the school achieved "exemplary" tests scores on the TAAS from the Texas Education Agency. The school outperformed not only the district averages at each grade level but also Texas state averages for all assessments of reading and mathematics except for those of fourth grade math. In addition, economically disadvantaged students at Sam Houston dramatically outperformed the comparison group of economically disadvantaged students across Texas at teach grade level. While all of the ferment on the campus, in the community, and in the relationship with Saint Joseph's the Worker were intrinsically important, it was an additional source of satisfaction to know that the work had also paid off in terms of the test results. As one might have anticipated given Coleman's early work in social capital theory, the social capital that had developed between the school and the community had created human capital in the form of students with academic skills and a real knowledge base.

The appendix provides an overview of Sam Houston's progress on TAAS tests from 1993 to 1999. If one begins with Table 1, which provides an overview of students' reading and mathematics scores, a number of trends are evident. Most noticeable is the strong upward trajectory in achievement from 1993 to 1998. Even when Sam Houston students began at a level that exceed both district and state scores – as was the case for third graders in both reading and mathematics – they generally continued their strong performances across those years. In 1998 all of the third, fourth, and fifth grade students outperformed their peers in the district and state in reading and mathematics, with the single exception of fourth graders in mathematics. As Table 2 indicates, this same high level of achievement also held for the fourth grade students on the writing section of the TAAS.

Aside from these felicitous findings, however, the data also reveal two major areas of concern. Table 3 shows the test results of those Sam Houston third graders who took the TAAS in Spanish from 1997 to 1999. In spite of Sam Houston's emphasis on cultural responsiveness, it is striking to note how much more poorly Sam Houston's students did on this instrument than other students in the same cohort at the district and the state level. More work in this domain, perhaps paying less attention to cultural influences and instead focusing more specifically on linguistic capabilities or test-taking skills, would appear to be warranted.

The second major area of concern relates to the slump in scores at the third and fifth grade levels from 1998 to 1999 in both reading and mathematics, but especially in reading. As Table 1 shows, only the fourth grade students out-performed their peers in reading and mathematics at the state and district level in 1999. As a consequence of these results, Sam Houston's rating by the Texas Education Agency fell from "exemplary" to "recognized" in 1999.

It is difficult to know why the scores developed as they have at Sam Houston. When queried, Connie Maheshwari was upbeat. "It means we have to work harder," she said. "It doesn't mean that our kids don't know the answers to the questions on the test. I think it's more likely that they might have been unfa-miliar with the format. Sometimes slight differences in wording or presentation can make a big difference for a child" (personal communication, June 9, 1999). Recognizing that test scores often fluctuate in the history of the school, Sam Houston has identified areas in which students performed poorly and is focusing its instructional strategies on those pupils who are struggling.

Alamo Middle School

Sam Houston Elementary School demonstrates the many positive developments that can occur when community organizing becomes a central part of the culture of a public school. It would be disingenuous, however, to suggest that all schools that enter the Alliance School network experience such fortuitous developments. One Alliance School with a more contested history is Alamo Middle School, located a few miles east of Palmer on Highway 83 in Alamo. Opened in 1988 to accommodate Alamo's growing population, the middle school was a tough place for students and teachers alike in its first few years. In spite of its sparkling new facilities, gangs dominated the hallways and teachers constantly had to discipline students for food fights, unruly behavior in classrooms, and alterca-tions in restrooms. Frightened by the breakdown in respect for authority, teachers withdrew from one another and focussed on their own classrooms. "One of the worst things of that time," one sixth-grade teacher remembered, "is that teachers wouldn't support other teachers. A student would cuss you out, or run down the hall, and even if other teachers saw it, they wouldn't say anything." The first principal, Scott Owings, struggled without success at Alamo for two years, at which point the school board decided that a change in lead-ership was necessary.

The school board thought that they found a possible leader for Alamo Middle School in René Ramirez, an assistant principal at Pharr-San Juan-Alamo High School, known in the area simply as "PSJA." Ramirez had been in charge of discipline at PSJA and had done an effective job. In addition, Ramirez seemed eager to be in charge of a campus, and to expand his leadership beyond simply keeping order in the school.

Recognizing the need for new information, Ramirez formed teams of teachers and parents in his second year that went with him to visit schools that were exploring innovative approaches to instruction, curriculum design, assessment, and community engagement in Texas. One school that they visited was Frank J. Dobie Middle School in Austin, which gave them a number of ideas about strategies for transforming the internal culture of Alamo from that of a junior high school to a more age-appropriate middle school concept. Dobie radicalized Ramirez's concept of what needed to be done with Alamo. "When I first got here," he said, "what we really had was a miniature high school. Our layout was just like a high school, with a bunch of different departments scattered around and no one really keeping track of our kids in a way that allowed for any oversight. Our meetings only involved department heads; most of the teachers weren't involved. The whole focus was on the content area.

"But we knew that wasn't working for us. We learned that there was a middle school concept which aimed to do something different: to focus totally on the student." To ensure that Dobie's success was replicable, Ramirez and his colleagues planned another site visit, to Berta Casava Middle School, in nearby San Benito. Just like Dobie, Berta Casava had teams of teachers working closely with students and meeting on a daily basis to discuss difficulties with students, curricular alignment, and all of the myriad problems which arise as part of the process of teaching young adolescents. Unlike Alamo, students appeared to be calm, focused on their studies, and respectful. "The whole school atmosphere was different," one teacher recalled. "We saw kids in the hallways who were unsupervised working on projects and they were all under control. They were getting things done and doing fine. At that time that was a foreign concept for us because we still had lots of gang problems. We were wowed!" (L. Whitlock, personal communication, November 17, 1997).

Emboldened by the success they witnessed at Dobie and Casava, Ramirez and his colleagues decided to take a risk in the fall of 1993 and to try the middle school structure at Alamo. For Ramirez, that meant shifting a focus to a more student-centered organization in which interdisciplinary clusters of teachers would work to get to know students well and would operate through a network of small teams in a common wing of the school. Yet rather than risk a tumultuous reorganization of the entire school at the outset, an experimental cluster would be initiated, and Ramirez found some volunteers who were willing to give it a chance. Among others, teachers Robert Martinez, Ermilia Sanchez, and Dervin Koenig agreed to pilot the concept with a group of roughly 150 students in common.

At roughly the same time that Ramirez and his colleagues made different site visitations, the principal was approached by his superintendent, Ernesto

Alvarado, who told him about a new opportunity to develop strong ties to Valley Interfaith and a new network of schools, called "Alliance Schools." Ramirez had developed a positive impression of Valley Interfaith and its work on colonias, get-out-the-vote drives, and health care issues. "I was so excited by my first contacts with Valley Interfaith," he recalled. "I wanted people in our community to understand what was happening inside our school. I wanted them to come and to ask us what we were doing for our migrant laborers, for our immigrants, and for our gifted and talented. We know that without a good parental involvement program in our school, we're not going to get the success that we need. I'd like to fill up the cafeteria with parents when we have PTO meetings. I envision parents in the classroom. I envision parents monitoring the hallways." Regrettably, when a relatively inexperienced community organizer made her first presentation on behalf of Valley Interfaith to Alamo faculty, many teachers felt threatened by the tone of urgency she brought to their meeting.

Elisabeth Valdez, another Valley Interfaith organizer, then tried a more exploratory and dialogical approach. She met with teachers and parents and discovered that in spite of the fact that the school enrolled roughly eight hundred students, meetings of the Parent-Teacher Organization usually were attended by no more than twenty parents. She stressed to the teachers that the idea of Alliance Schools was not to force change on reticent educators, but to develop positive relationships with the community which could make their work easier and more effective. Finally, Valdez invited Alamo teachers to attend state-wide meetings of the Alliance Schools; at those meetings teachers learned that a whole network of schools working with community-based organizations was developing throughout Texas, and was doing so in a manner which stressed accountability and the building of broad networks of civic engagement. Within the Rio Grande Valley, schools in Brownsville, Pharr, McAllen, and a number of the smaller cities between them were also coming on board. While most of the schools joining the network were elementary schools, Travis Middle School in McAllen had become the first secondary school in the Valley to become an Alliance School. Thanks to Valdez's tact and Ramirez's support, the idea of developing Alamo into an Alliance School became attractive once more to the teachers.

Once the teachers' efforts began garnering increased parental engagement with Alamo Middle School, they agreed formally to become an Alliance School. One of the first and most palpable benefits of this decision was additional funding. Although the fifteen thousand dollars that were provided in the first year of funding might not seem like much when viewed from the perspective of a large corporation, for a school in a small and financially strapped district it enabled teachers to participate in valuable professional development work-shops. In addition, Alliance School meetings with guest speakers from leading research universities and smaller gatherings with principals from similar schools in low-income communities around the state enabled Ramirez and his colleagues to gain new ideas from colleagues struggling with similar problems. Perhaps most important, however, was the larger cultural change entailed in shifting teachers' attention from curriculum and assessment to the broader identifies and concerns of the students they taught. "As an Alliance School we focus on the whole student," Ramirez said, "not just the individual that happens to be within our walls. We look at the whole environment – the streets in our community, health issues, jobs, crime – and try to understand what this means for our students. And we know that there are changes that need to happen not only in our school but also in our community" (personal communication, May 22, 1997).

Simultaneous with the initiation of the Alliance School partnership at Alamo, René Ramirez and his colleagues evaluated the development of the first teacher team in Alamo. There were many problems that first year. Each of the teachers on the team had inordinately large classes – Darvin Koenig remembered that all of them had roughly thirty-six students – and because teachers on the team had time for a "team conference" in addition to their "personal conference," other teachers resented what they construed as an arbitrary perk. A further point of jealousy among many staff was that, other than Martinez, all of the teachers on the team were young and comparatively inexperienced; some veteran teachers felt that they had been unfairly passed over and would have gladly welcomed two conferences in their work day. In spite of the problems, however, the first teacher team reported enthusiastic results with its students, and other teachers and students were eager to give it a try as well.

In the ensuing years Alamo developed ten teams of teachers, and teachers and students worked together to find creative names for their learning commu-nities; some of the finalists were the "Starcatchers," the "Killer Bees," the "Determinators," the "Dream Team," the "Challengers," the "Texas Tornadoes," and the "Rainbow Riders." To strengthen a sense of group identity that is so important for middle school students, teacher and student teams have devel-oped logos, their own T-shirts, and even their own stationary with team letterhead. In addition, they have developed a lively sense of competition that is translated into athletic events, school festivals, and friendly banter with one another. "That change was the best thing that ever happened in our school," Ramirez enthused, and noted with pride that the three other middle schools in his district all followed Alamo's lead in this regard in subsequent years (R. Ramirez, personal communication, May 22, 1997).

In the summer of 1995, Ramirez was appointed to become the principal of Pharr-San Juan-Alamo High School known locally simply as "PSJA." He was replaced by Rosie Ruiz, who had been an assistant principal at the school since its inception. Like most other civic leaders in the Valley, Ruiz is an immigrant, who was born in the small village of Hualahaises in Nuevo Léon and moved to the United States with her parents when she was one year old. She attended the public schools in Pharr and Alamo, and graduated from PSJA in 1970. She graduated from Pan American University in three years and returned to PSJA as a teacher in 1973, where she taught English and Spanish for fourteen years. Then followed a one year break to work for the district, coordinating language arts instruction at the middle school level; she visited three campuses to monitor and improve the instruction of English teachers. She then was hired as an assistant principal at the brand new Alamo Middle School that opened in the fall of 1988.

Rosie Ruiz inherited a school that had developed momentum under the leadership of René Ramirez, but she was determined to place her own imprint on the building. The major reforms that have occured under her leadership are to make a transition to longer class periods in a reform known as "block scheduling" and to implement a process of peer mediation in which students learn how to intervene and resolve conflicts between themselves. The transition to block scheduling has been relatively minor in the school. In most of the teams, teachers use an "alternate" form of block scheduling, in which different classes are held on Tuesdays and Thursdays than on Mondays, Wednesdays, and Fridays. Yet it is intriguing to note that teams have abundant freedom to experiment with their own structures and that the collaborative structure seems to enhance teacher autonomy rather than to constrain it. The Stargazers, for example, use an accelerated form of block scheduling in which team members teach the same students every day for three weeks; they then rotate the classes around to each other. In addition, the Stargazers decided that they wanted to keep the same students for two years in a row, which team members enjoy because of their ability to trace students' progress over a longer period of time. Somewhat like a federal system in which states are interdependent in some areas and autonomous in others, Alamo has developed a culture in which teams of teachers can develop their own professional judgment about the best way to educate their pupils.

Many of the changes described above occurred primarily as a result of internal developments in Alamo. Yet what of the school's collaboration with Valley Interfaith? Its presence at Alamo has continued to be contested throughout the history of the collaboration. The first presentation by Valley Interfaith at Alamo provoked much anxiety and resistance among teachers to the idea of a collaboration with the community organization, and although Elizabeth Valdez was able to repair that, subsequent difficulties have ensued. Industrial Areas Foundation organizers, in line with their principle of the "iron rule" – the notion that one should never do for others what they are capable of doing for them-selves – often move on to another site quickly after organizing one school. This appears to have been the case at Alamo, where teachers felt at a loss after encounters with Valley Interfaith that appeared brief and inconsequential.

The diminished presence of Valley Interfaith on the Alamo campus produced grumbling among teachers on the campus in the following years, and points to a particular tension that accompanies the actualization of the iron rule. For while some educators are pleased to know that IAF organizations plan to teach a new kind of community engagement and then to depart, respecting the capacity of educators to develop social capital as they best see fit, other teachers seek a more sustained relationship and guidance. "When I voted to partner with Valley Interfaith, I thought that I was voting for them to really be here on campus to help us to organize our relationships with the community better," one teacher complained. "Instead, they only seem to show up when they have a meeting coming up and want our participation." Teachers on one team felt clearly manipulated when they were asked by an assistant principal at Alamo to support Valley Interfaith's "sign up and take charge" campaign of voter registration in September 1997. "This really isn't part of our job," one seasoned veteran complained, "and I don't see why we should get out and support Valley Interfaith's agenda when we haven't seen hide nor hair of them for a long time now."

Teachers' resistance to Valley Interfaith reveals a point of friction where professional autonomy and community engagement are not easily combined. On the one hand, when teachers perceive a lack of evidence of support by Valley Interfaith for their school, this interpretation can aptly be criticized for a certain kind of selective perception, for teachers would likely be upset to lose the programs that Alliance School funds support, such as peer mediation, parent education classes, and professional development for teachers. On the other hand, the teachers' resentment in this particular case stemmed from a sense that Valley Interfaith was incurring on their professional autonomy. The fact that an assis-tant principal asked the teachers to support Valley Interfaith's agenda may also be seen as reflective of a schism between administration and faculty who are reluctant to follow the administration's politics, even if Valley Interfaith claims to be non-partisan.

It wasn't until two months later, in November 1997, that Valley Interfaith organizer Estela Sosa-Garza attended a faculty meeting at Alamo. At that time she sought to elicit faculty support for an evening event which would welcome parents into the school and ask them in small group settings what they like or don't like about Alamo. As one might anticipate given the context, some teachers vocally expressed skepticism. One teacher was concerned that the meetings would be "superficial" and would fail to impress upon the parents just how weak their children's academic work was and the need for full parental support in the effort to bring their children's work up to grade level. "I don't think we need to ask parents what they like or don't like about our school, like Valley Interfaith says," he commented, "I think we need to let the parents know that their kids achieving is at a level far lower than what they're capable of. We need to give the parents a wake-up call." He was supported by a colleague, who remarked, "I grew up in the colonias in San Juan and I know what the conditions there are. I don't need to hear about that. But the minute our kids get home, the parents speak to them only in Spanish, they watch Univision, and everything that we're trying to do here gets washed away. We take one step forward here, and the kids go home, and it's two steps back. We need to tell the parents that they need to be much more active to help their kids to succeed or the whole cycle is just going to be perpetuated all over again."

It was an auspicious moment for Alamo, where the principals of the Alliance School network met head-on with the wariness of hard-working veteran teachers. Yet it is important to note that no teachers spoke out against greater communication between the school and the parents; rather, the *nature* of that communication was questioned. Valley Interfaith, sensitive to the intimidation that many low-income parents feel when entering a school, sought to win the parents' trust and to engage their critical thinking by creating a forum in which parents would feel that the school genuinely cared about their perceptions and sentiments. Oppositional teachers, on the other hand, felt it imperative that the school impress upon the parents both the need to reorient everyday household cultural patterns to promote greater academic success as well as the need to be forthright about students' level of academic ability.

Alamo Middle School resolved the disagreements among its faculty briefly by hosting an open house for parents which had high turn out and improved staff morale. Soon enough thereafter, however, the critical questions about strategies for engaging parents re-emerged, and they have persisted up to the time of this writing. Unlike Sam Houston, Alamo has not succeeded in engaging parents who will teach classes, conduct house meetings on their own initiative, and develop a broader educational improvement strategy in collaboration with a community organization like Valley Interfaith.

When one turns to TAAS data for Alamo students, the results indicate steady improvement over the years. A new TAAS test was implemented in Texas schools during the 1992-1993 school year, so this account begins with that data.

In the first administrations of the new TAAS in 1993 and 1994 Alamo students were struggling. As Table 4 in the Appendix indicates, sixth graders performed worse than their peers at the district and the state level, and disadvantaged Alamo sixth graders performed worse than their peers at the state level. At the seventh grade level, Alamo students outperformed their peers in the district, but not the state, and there was little difference between disadvantaged students and their peers at the state level in reading and mathematics. Eighth graders likewise outperformed the district, and lagged behind the state; however, disadvantaged students performed worse than their peers at the state level. Eighth grade writing scores are shown in Table 5, and they exhibit that the general pattern suggested above continues: Alamo students performed better than the district and worse than the state. In this instance, however, disadvantaged Alamo students outperformed their peers at the state level. This achievement should be relativized, however, since more than a third of those students failed to pass the writing section.

From these inauspicious beginnings Alamo's TAAS results have shown a pattern of steady improvement in subsequent years. The rate of improvement of sixth grade students in reading and math from 1993 to 1999 surpasses that of both the state and the district, and if present trends continue, Alamo will overtake the state and district scores in the next few years. Seventh grade reading scores have been modest, as have been the gains at the state and district level, but the mathematics gains have been dramatic (almost 30% from 1994 to 1999). Eighth grade reading scores have improved steadily and we can extrapolate that the school will surpass the state if present trends continue. Mathematics gains have been so dramatic that Alamo students outperform not only district students but also Texas students, with disadvantaged students at Alamo outperform disadvantaged students across the state by a wide margin. As Table 5 shows, Alamo eighth graders now outperform at the state and district on the writing section of the TAAS as well.

Alamo would thus appear to be on a strong upward trajectory in terms of academic achievement at the sixth and eighth grade levels, with a need for additional attention to improvement at the seventh grade. Many faculty are pleased with their students' achievement on the test. However, it is also important to note that many are angry about the emphasis on the TAAS in the school and frustrated by their students' continued difficulties on the test. "We have to remember that the TAAS is a *minimum* skills test," one teacher commented, "and we're still not excelling even on that measure." Other teachers worry that so much attention is focused on the TAAS in the school that students' higher-level, critical thinking skills are rarely engaged. "If I just taught to the test, I'd completely lose my kids," one English teacher commented, "and besides, it wouldn't even be right. How are you ever going to get a love of literature or develop any complexity in your thinking by cramming for the TAAS?"

While one should appreciate Alamo's gains in student achievement on the TAAS, one should also recognize the costs paid by teachers and students to reap those results. The chair of the school's language arts program, Diane Hinojosa, took twelve-minute lunch breaks from August through January of the 1998-1999 school year so that she could intensely tutor over fifty students over the lunch period to boost their TAAS writing scores. When the results came in, 95% of her students had passed. She cried and so did many of her students and colleagues. "They really weren't sure that they could do it," she recalled, "and when the results came in it was such a victory for them." The poignant part of this story is that teachers and students deserve lunch breaks, and while it is gratifying to know that the students passed, concern about the manner in which standardized tests are driving teachers and students to abandon hitherto sacrosanct times to catch one's breath and reflect on the morning's events seems appropriate (D. Hinojosa, personal communication, August 25, 1999).

By the fall of 1999 Alamo Middle School was focusing its efforts on TAAS preparation and a new student discipline plan, and the relationship with Valley Interfaith was in trouble. The small group of faculty who truly believed in the importance of community engagement were exhausted by the constant rounds of meetings that were called by Valley Interfaith which they felt bore meager results. "I've got two kids at home," one teacher complained, "and it's not right for me to be coming home at nine o'clock, long after they've gone to bed, or to be out at workshops all day on Saturdays." "As far as I can tell, the Alliance has fizzled out," another teacher commented. "It was good when we were first getting started, because it really did get us together with the parents and the community. But now we're in a different place. We've internalized all of that and need to move on." Reflecting the troubled nature of the relationship with Valley Interfaith, Alamo faculty voted against even applying for a renewal of Alliance School funds in November 1999, although the school still participates in the Alliance School network.

Interpretation

Americans are fond of seeking panaceas for our troubled public schools. Our current interest in the role that community-based organizations can play in improving schools is warranted, given the problems with the many more orthodox school reform efforts that have fallen flat over the last quarter century. A reading of these two case studies from South Texas, however, indicates that community-organizations have many challenges of their own in building strong lateral ties to public schools.

Alamo Middle School illustrates the many obstacles to social capital formation between schools and community organizations. Oppositional teachers at Alamo doubted the efficacy of Valley Interfaith's strategy for school and community improvement. The relationship began with teacher opposition, evolved to sustain a measure of collaboration, and then fell back to a condition of skepticism and immobility among Alamo's faculty. Alamo's primary innovations in the 1990s – such as cross-disciplinary teams and block scheduling – occurred with a minimum of community engagement and represent fairly orthodox, school-based approaches to educational change. Parents played virtually no role in informing curricular decisions, and some oppositional teachers clearly saw parents as obstacles with cultural practices that impede student learning.

Sam Houston Elementary School, on the other hand, indicates the obverse side of the coin. Sam Houston teachers and parents worked with Saint Joseph's the Worker Catholic Church to promote a Kids' Action Assembly that held public officials accountable for the conditions in their neighborhood. Sam Houston teachers and parents then used the energy derived from this assembly to acquire a new school, develop culturally responsive curricula, and engage parents in a wide range of instructional units. Sam Houston teachers spoke at length about the benefits of parental engagement and the importance of developing relationships with local church and civic leaders. If Alamo shows just how difficult school and community collaborations can be, Sam Houston indicates how rich the rewards are when mistrust and skepticism are overcome in the interests of a common commitment to children and their community.

What were the conditions that enabled Sam Houston to benefit from its collaboration while Alamo faltered? Sam Houston manifested strong *bridging* social capital with Saint Joseph's the Worker Catholic Church. Saint Joseph's had first organized La Paloma and other barrios within its parish boundaries, and Sam Houston skillfully capitalized on this prior work when its own parental engagement strategy was under development. Sam Houston teachers and parents united themselves by recognizing the many problems in their community that impeded children's learning, and by developing a common strategy to attack those problems directly. Through their conjoint efforts to improve their community and their school, Sam Houston teachers and parents developed a mutual respect for one another that was grounded in the educational aspirations of the the Sam Houston community for the children.

Alamo Middle School faculty had an intellectual awareness of the need for strong community ties, but their academic commitments ironically undermined their efforts to make the most of the social capital in the community. Alamo's oppositional teachers manifested a single-minded loyalty to their disciplines and academic achievement – a fairly traditional human capital approach, focussed on the learning gains of individuals, typically as measured in the state standardized tests. Alamo teachers appear to have overlooked the manner in which strong community ties can support achievement rather than distract from it.

The test score results indicate that the emphasis on human rather than social capital has not been without some gains for Alamo. Particularly at the sixth and eighth grade levels, Alamo students have improved at impressive rates on the TAAS. On the other hand, one also notes that Alamo has not achieved the same "exemplary" rating on the TAAS as Sam Houston. Recognizing that the findings from individual school cases must be interpreted cautiously, it does appear that Sam Houston's mixture of social and human capital is a more effective school improvement strategy than a concentration on human capital alone.

ACKNOWLEDGMENTS

The author wishes to acknowledge the support of funding for this research from the Annenberg Rural Challenge, Boston College, the Mellon Foundation, the Ford Foundation, and the American Academy of Arts and Sciences. In addition, I would like to thank Steven Blaum, Brionne Chai-Onn, Aaron Ramirez, and Kate Sorgi, my research assistants at the Lynch School of Education at Boston College for their help with data collection and analysis. Finally, I would like to thank Mary Brabeck, Oralia Garza-Cortés, Jim Fleming, Walt Haney, Cathy Horn, Brinton Lykes, Guadalupe San Miguel, Angela Valenzuela, and Emilio Zamora for their invaluable comments on this research.

REFERENCES

Becker, G. S. (1964). *Human Capital: A Theoretical and Empirical Analysis, with Special Reference to Education*. New York: National Bureau of Economic Research.
Bryk, A. S., Lee, V. E., & Holland, P. B. (1993). *Catholic Schools and the Common Good* (pp. 306–308, 378). Cambridge, Mass.: Harvard University Press.
Cohen, J. L., & Arato, A. (1992). *Civil Society and Political Theory*. Cambridge, Mass.: MIT University Press.
Coleman, J. S. (1988). Social Capital in the Formation of Human Capital. *American Journal of Sociology, 94*(Suppl.), S95–S120.
Coleman, J. S. (1990). *Foundations of Social Theory* (pp. 300–321). Cambridge: Harvard University Press.

Coleman, J. S., Hoffer, T., & Kilgore, S. (1982). *High School Achievement: Public, Catholic, and Private Schools Compared*. New York: Basic Books.

Coleman, J. S., & Hoffer, T (1987). *Public and Private High Schools: The Impact of Communities*. New York: Basic Books.

Greeley, A. (1997). Coleman Revisited: Religious Structures as Sources of Social Capital. *American Behavioral Scientist, 40*(5), 587–594.

Hall, J. A. (Ed.) (1995). *Civil Society: Theory, History, Comparison*. Cambridge, Mass.: Polity.

Hargreaves, A. (1994). *Changing Teachers, Changing Times: Teachers' Work and Culture in the Postmodern Age*. New York: Teachers College Press.

Herrera, D. (1999, August 18). Alamo Middle School Launches Discipline Program. *The Advance News Journal*.

King, B. (1994, April 1). Sixth Graders Will Stay at Alamo Middle School. *The Monitor*.

Loury, G. (1977). A Dynamic Theory of Racial Income Differences. In: P. A. Wallace & L. Le Mund (Eds), *Women, Minorities, and Employment Discrimination*. Lexington, Mass.: Lexington Books.

Loury, G. (1987). Why Should We Care About Group Inequality? *Social Philosophy and Policy, 5*, 249–271.

Muncey, D. E., & McQuillan, P. J. (1996). *Reform and Resistance in Schools and Classrooms: An Ethnographic Portrait of the Coalition of Essential Schools*. New Haven: Yale University Press.

Nava, J. D. (1999, August 29). School Has New Discipline Program. *The Monitor*.

Portes, A. (1998). Social Capital: Its Origins and Applications in Modern Sociology. *Annual Reviews in Sociology, 24*, 1–24.

Putnam, R. (1995). Bowling Alone: America's Declining Social Capital. *Journal of Democracy, 6*(1), 65–78.

Putnam, R. (1993). *Making Democracy Work: Civic Traditions in Modern Italy*. Princeton: Princeton University Press.

Sarason, S. (1990). *The Predictable Failure of School Reform*. San Fransisco: Jossey-Bass.

Shirley, D. (1997). *Community Organizing for Urban School Reform*. Austin, TX: University of Texas Press.

APPENDIX

Table 1. Percentages of Students Passing the Reading and Math Sections of the TAAS, Sam Houston Elementary School, 1993–1999.

		State	District	Campus	Econ. Disadv.(School)	Econ. Disadv. (State)
Grade 3						
Reading	1999	88.0%	89.0%	82.8.0%	80.7%	81.6%
	1998	86.2%	83.3%	100.0%	100.0%	79.0%
	1997	81.5%	75.0%	80.0%	78.0%	72.0%
	1996	80.5%	80.8%	77.8%	76.5%	70.1%
	1995	79.5%	78.2%	58.3%	59.4%	68.7%
	1994	77.9%	83.3%	91.5%	90.2%	66.4%
Math	1999	83.1%	84.4%	87.7%	86.2%	75.1%
	1998	81.0%	78.3%	92.3%	91.4%	72.2%
	1997	81.7%	79.5%	82.2%	80.5%	73.3%
	1996	76.7%	78.6%	77.8%	76.5%	66.4%
	1995	73.3%	69.8%	35.1%	36.4%	61.8%
	1994	63.0%	62.0%	85.1%	85.4%	49.1%
Grade 4						
Reading	1999	88.8%	91.5%	96.0%	95.2%	82.3%
	1998	89.7%	89.7%	96.2%	96.0%	83.4%
	1997	82.5%	84.3%	86.7%	86.0%	73.0%
	1996	78.3%	82.9%	82.9%	84.4%	67.5%
	1995	80.1%	86.8%	95.1%	94.3%	69.2%
	1994	75.5%	82.5%	69.0%	68.3%	63.3%
	1993	65.5%	69.8%	64.3%	63.4%	49.4%
Math	1999	87.6%	91.4%	98.1%	97.7%	81.3%
	1998	86.3%	88.9%	84.6%	84.0%	79.5%
	1997	82.6%	85.8%	88.9%	90.7%	73.9%
	1996	78.5%	81.9%	82.9%	84.4%	68.3%
	1995	71.1%	74.1%	85.4%	82.9%	58.2%
	1994	59.4%	66.6%	66.7%	65.9%	45.7%
	1993	52.6%	52.9%	54.8%	53.7%	36.9%
Grade 5						
Reading	1999	86.4%	87.8%	80.4%	78.8%	78.0%
	1998	88.4%	91.5%	100.0%	100.0%	81.7%
	1997	84.8%	87.7%	92.5%	92.3%	75.7%
	1996	83.0%	89.3%	85.4%	83.8%	73.1%
	1995	79.3%	84.7%	686.0%	85.7%	68.4%
	1994	77.5%	86.4%	88.6%	88.4%	65.4%

Table 1.   Continued.

		State	District	Campus	Econ. Disadv.(School)	Econ. Disadv. (State)
Math	1999	90.1%	92.7%	89.1%	88.2%	84.9%
	1998	89.6%	93.3%	96.0%	97.9%	84.0%
	1997	86.2%	90.6%	89.7%	89.5%	78.7%
	1996	79.0%	88.3%	92.7%	91.9%	68.7%
	1995	72.6%	78.1%	88.6%	88.4%	60.2%
	1994	62.6%	73.3%	58.1%	57.1%	48.4%

Table 2.   Percentages of Fourth Graders Passing the Writing Section of the TAAS, Sam Houston Elementary School, 1993–1999.

		State	District	Campus	Econ. Disadv.(School)	Econ. Disadv. (State)
Grade 4						
Writing	1999	88.4%	90.3%	88.0%	90.2%	83.3%
	1998	88.7%	88.5%	90.4%	90.0%	83.0%
	1997	87.1%	86.6%	91.1%	90.7%	80.4%
	1996	86.3%	88.0%	80.0%	78.1%	79.9%
	1995	85.0%	88.8%	97.6%	97.1%	77.2%
	1994	85.5%	91.3%	85.0%	84.6%	77.4%
	1993	83.4%	88.0%	86.0%	85.7%	73.8%

Table 3.   Percentages of Third Grade Students Passing Spanish-Language Administrations of the TAAS in Reading and Mathematics, Sam Houston Elementary School, 1997–1999.

		State	District	Campus	Econ. Disadv. (School)	Econ. Disadv. (State)
Spanish Grade 3						
Reading	1999	74.2%	88.9%	84.0%	82.0%	73.8%
	1998	65.6%	50.0%	23.5%	23.5%	65.4%
	1997	44.6%	38.1%	23.1%	23.1%	44.3%
Math	1999	74.9%	90.0%	NA	NA	74.6%
	1998	66.4%	53.1%	23.5%	23.5%	66.3%
	1997	53.5%	40.5%	15.4%	15.4%	53.2%

Table 4. Percentages of Students Passing the Reading and Math Sections of the TAAS, Alamo Middle School, 1993–1999.

		State	District	Campus	Econ. Disadv.(School)	Econ. Disadv. (State)
Grade 6						
Reading	1999	84.9%	79.4%	80.8%	79.9%	76.1%
	1998	85.6%	71.9%	75.7%	73.3%	75.8%
	1997	84.6%	70.9%	73.9%	71.9%	74.3%
	1996	78.4%	55.9%	59.8%	58.6%	64.6%
	1995	78.9%	65.7%	63.6%	59.7%	66.7%
	1994	74.1%	61.5%	55.6%	51.0%	60.1%
Math	1999	86.9%	89.0%	89.1%	89.2%	80.2%
	1998	86.1%	83.3%	82.7%	81.7%	78.4%
	1997	81.8%	77.3%	78.0%	76.3%	71.7%
	1996	77.8%	66.5%	75.7%	76.1%	66.1%
	1995	64.6%	51.0%	43.2%	37.6%	48.6%
	1994	61.1%	55.2%	41.8%	39.4%	45.4%
Grade 7						
Reading	1999	83.6%	72.6%	73.2%	70.4%	73.6%
	1998	85.5%	73.2%	76.2%	74.9%	75.5%
	1997	84.5%	65.5%	69.3%	68.8%	74.0%
	1996	82.6%	67.6%	65.5%	62.4%	71.4%
	1995	78.7%	64.6%	64.2%	61.2%	65.9%
	1994	75.9%	61.0%	66.3%	62.5%	61.0%
Math	1999	84.9%	80.5%	77.4%	75.2%	77.0%
	1998	83.7%	78.9%	80.3%	79.3%	73.7%
	1997	79.7%	71.8%	75.6%	74.6%	68.8%
	1996	71.5%	60.6%	61.9%	58.8%	56.6%
	1995	62.3%	47.7%	45.8%	43.3%	44.5%
	1994	59.7%	43.8%	47.5%	44.4%	42.2%
Grade 8						
Reading	1999	88.2%	79.7%	85.7%	84.1%	80.7%
	1998	85.3%	69.0%	74.0%	71.6%	74.8%
	1997	83.9%	69.0%	73.6%	70.4%	72.7%
	1996	78.3%	64.7%	64.3%	60.5%	64.3%
	1995	75.5%	61.9%	62.8%	58.4%	60.5%
	1994	77.2%	61.0%	68.5%	65.5%	61.9%
	1993	71.8%	46.0%	53.3%	47.0%	53.4%

Table 4. Continued.

		State	District	Campus	Econ. Disadv.(School)	Econ. Disadv. (State)
Math	1999	86.3%	84.3%	88.6%	89.3%	78.7%
	1998	83.8%	77.2%	79.8%	78.7%	74.6%
	1997	76.3%	67.0%	70.8%	68.1%	63.6%
	1996	69.0%	58.8%	56.9%	55.1%	53.4%
	1995	57.3%	39.3%	33.7%	31.3%	37.8%
	1994	58.6%	42.1%	47.7%	45.9%	39.9%
	1993	51.1%	26.5%	30.1%	25.5%	30.1%

Table 5. Percentages of Eighth Grade Students Passing the Writing Section of the TAAS, Alamo Middle School, 1993–1999.

		State	District	Campus	Econ. Disadv.(School)	Econ. Disadv. (State)
Grade 8						
Writing	1999	85.7%	83.1%	88.9%	87.1%	77.6%
	1998	84.0%	74.8%	84.1%	82.6%	74.7%
	1997	80.7%	70.5%	73.5%	70.4%	69.4%
	1996	76.9%	67.6%	67.5%	66.2%	63.8%
	1995	75.3%	59.9%	63.3%	60.8%	62.2%
	1994	69.8%	59.7%	62.9%	59.1%	55.0%
	1993	73.9%	60.5%	67.1%	63.1%	58.9%

Table 6. Percentages of Sixth Grade Students Passing the Spanish-Language TAAS in Reading and Mathematics, Alamo Middle School, 1998–1999.

		State	District	Campus	Econ. Disadv.(School)	Econ. Disadv. (State)
TAAS % Passing Spanish Grade 6						
Reading	1999	30.2%	NA	NA	NA	30.2%
	1998	28.2%	41.2%	41.2%	37.5%	28.0%
Math	1999	51.2%	NA	NA	NA	52.1%
	1998	38.2%	33.3%	33.3%	29.4%	38.4%

LESSONS (AND QUESTIONS) FROM WORKPLACE SCHOOLS ON THE INTERDEPENDENCE OF FAMILY, SCHOOL AND WORK

Claire Smrekar

INTRODUCTION

Corporate-sponsored child care designed to better serve the family needs of employees has expanded recently to include elementary schools at the workplace. Over 40 companies, including Target Corporation, Bank of America, and Radisson Hotels, have established on-site elementary schools (Miller, 1997). The first workplace school was founded by the American Bankers Insurance Group in Miami, ten years ago. The arrangements require corporate partners to provide the facility and assume full responsibility for maintaining it; school districts provide the staff and assume full responsibility for instruction. These "schools of choice" give parents who are employed by the corporate sponsor the option of selecting the workplace for their children's public education.

The concept of workplace schooling appeals to employers for many of the same reasons that corporate child care has become popular and widespread. Benefits include: a decrease in absenteeism, tardiness, and turnover; an increase in employee performance, morale, and satisfaction and; an enhanced corporate reputation and recruitment ability as a family-friendly work environment (Munk,

Community Development and School Reform, Volume 5, pages 171–192.
Copyright © 2001 by Elsevier Science Ltd.
All rights of reproduction in any form reserved.
ISBN: 0-7623-0779-X

1996). Workplace schools are also attractive to educators. First, these public school-corporate partnerships may relieve overcrowding and reduce fiscal constraints in school districts with exploding enrollments (Murray, 1998). Second, in addition to the economic benefits, the idea of locating a public school close to parents' employment fits well with the interest in promoting parental involvement in education.

Workplace schools are expected to multiply over the next decade due to increasing corporate demand and parent-employee interest in these public schools of choice (Murray, 1998). Although financial analysts and business writers (see Cowans, 1997; Munk, 1996) have examined the corporate consequences of these public-private partnerships (e.g. return on investment), no studies to date have explored the concepts of social integration policy that are anchored to the educational experiences of parents, students, and teachers in these new public schools.

This paper explores the compelling questions and emerging findings from a two-year research project, funded by The Spencer Foundation, that focuses upon the nature and inter-related functions of work, family, school and "neighborhood." I argue that a set of important social trends are reflected in workplace schools, including an increased time spent at the workplace, a diminished sense of neighborhood community, and an increasing tension between education as a private benefit and education as a public good. These issues establish the crosshairs for a debate targeted on core democratic values, against the backdrop of the pressing priorities and competing demands of work and family life for American school parents.

WORK LIVES AND FAMILY STRUCTURES

Almost half of the American work force is comprised of two-parent working families (Galinsky, Bond & Friedman, 1993). These parent-employees are part of a work force that is spending more and more hours each month at work, taking fewer unpaid leaves, and enjoying shorter vacations each year (Bureau of Labor Statistics, 1989; Schor, 1991). In an American "culture called to work," married couples today spend 185 hours more per year at work than couples spent just 10 years ago. When compared to workers in other industrialized nations, Americans contradict a worldwide trend of *fewer* hours spent at work. A recent report by the International Labor Organization (1999) indicated that in contrast to workers in Japan, Korea, and Western Europe, the number of work hours clocked by Americans has steadily *increased* over the past two decades. American workers now spend almost two weeks more at work than their Japanese counterparts (ILO, 1999).

These trends have triggered increased attention to the growing phenomenon that University of California sociologist Arlie Russell Hochschild (1997) calls "the time bind," the temporal squeeze between work demands and family obligations during which the workplace becomes a primary place of residence and the essential object of adults' emotional and intellectual commitment. A national study conducted by the Families and Work Institute indicates that men spend an average of 49 hours a week at work; women average 42 (Galinsky, Bond & Friedman, 1993). At the same time, over 40% of adult Americans say they have too little time to spend with their families (Gallup & Newport, 1990). These work patterns are reflected in the relocation of adults' primary social networks from the religious, civic, and social organizations that marked earlier decades, to the place of work in today's society (Hochschild, 1997). In a recent survey of employees from Fortune 500 companies, almost half of all respondents identified work as "the place where they had the most friends;" only 16% identified their neighborhood; just 6% indicated a church or temple (Murray, 1998). Indeed, Harvard University political scientist Robert D. Putnam (1995) points to the pervasive social phenomenon of "bowling alone" as a vivid illustration of the loss of communal bonds and associational life – outside of work – across all segments of American society. Putnam's theme of "social decapitalization" refers to the 25% decline in the membership rolls of social and religious groups, service and civic organizations (including the PTA, which declined from 12 million in 1964 to 7 million in 1995), and the attendant loss of social trust, shared norms and networks that define social capital (see Coleman, 1987). Although Putnam blames "the technological transformation of leisure" (i.e. television) for these patterns of social isolation and disconnection, the U.S. Census Bureau General Social Survey responses over the last two decades indicate the increasing strength of workplace-based social ties in lieu of neighborhood-based connections. These social changes and demographic trends raise new issues for educators, policymakers, and employers: Is the workplace the new American neighborhood?

CENTRAL RESEARCH QUESTIONS IN THIS STUDY

In earlier work (Smrekar, 1996), I examined the impact of cultural capital (see Bourdieu, 1977; Lareau, 1989), including the structure of work lives, on patterns of parental participation across choice and non-choice schools. The findings from these multi-case studies of a neighborhood school, a magnet school, and a Catholic school underscore the dominant influence of parents' work lives on patterns of participation in their children's schooling:

> Across social class and school setting . . . for many parents, the exigencies of work and family lives demand deliberate rationing of their limited time and energy. School-based

activities such as PTA meetings, spaghetti dinner fundraisers, and parenting workshops are considered extras which they cannot afford. Unless their children are involved in an event or activity, there is little interest in making the effort. Even then, many parents find it difficult to adjust their work schedules to fit school hours (Smrekar, 1996, p. 143).

This work argues for rethinking the social organization of schools and the work structures that prevent many parents from participating in parent involvement activities and in enhancing the educational experiences of children. This earlier study also provides a useful lens for examining public school policies that appear to be designed to avoid the conflicts around work lives, family structures and school organization. The Spencer Foundation-funded research study builds upon this and other previously published work (see below) by focusing upon the interaction between social structures (family life, work environments) and school organization.

By bridging the geographical barrier between work and school, the concept of workplace schools confronts the tension between a socially sanctioned pattern of increased time spent at work, and the increased demand for parental involvement in the educational lives of schoolchildren. The concepts of social capital and community provide the analytical frameworks for examining workplace school arrangements. Do workplace schools create enduring social ties among residentially/geographically dispersed families? Are the human capital resources available in corporate environments activated in workplace schools to enhance social capital? Are social structures in workplace schools reconstituted in ways that facilitate collective, purposeful action similar to magnet school communities (Smrekar, 1996) and Catholic school communities (Coleman, 1988)?

CONCEPTUAL FRAMEWORK

Social Capital

Amidst the increasing signs of social and residential isolation and racial segregation in America (Orfield, Bachmeier, James & Eitle, 1997), some scholars have suggested that the social ties that once connected Americans have violently eroded over the past decade in a convulsion of crass consumerism, elevated self-interest, and individualism (see Bellah, Madsen, Sullivan, Swidler & Tipton, 1985; Elshtain, 1995; Lasch, 1995). According to this view, the perceived decline of communal associations, including such social institutions as families, churches, unions, and civic groups, has silenced the political discourse that enjoins individuals of different ethnic, class, and religious lines to common action. The spirit and activities of Americans, notes political scientist Jean Elshtain (1995), suggest a "culture of distrust" that displaces a sense of shared interests, collective commitments, and mutual interdependence.

Against the backdrop of marked evidence of a diminishing sense of neighborhood community and involvement in communal associations (Putnam, 1995), workplace schools may represent new possibilities (while raising new concerns about access and equity) for a redesigned American neighborhood – at work – one stocked with ample, untapped social capital resources. Coleman (1987) defines social capital as "the norms, the social networks, and the relationships between adults and children that are of value for the child's growing up" (p. 36). The critical elements of social capital include shared values, norms, and attitudes that help promote trust, facilitate open and fluid communication, and produce purposeful and meaningful activities that benefit students and adults alike. Social capital is sustained when there is "a sense of community" or a set of organizational and institutional affiliations (e.g. civic, religious, professional) that bind families in stable, predictable, and enduring social ties (Putnam, 1995). Social capital bridges human capital theory (Schultz, 1963), which underscores the economic value of individuals for collective purposes, and social organization theory.

School Community

Contemporary concepts of community (e.g. Bronfenbrenner, Moen & Garbarino, 1984; Coleman & Hoffer, 1987; Newmann & Oliver, 1968; Scherer, 1972) assert that the community of residence in society today does not reflect the community of psychological meaning for most families (Steinberg, 1989). These observers distinguish between a concept associated with physical or geographical boundaries, and a concept of community grounded in social structures and social relations. For example, Newmann and Oliver (1968) include the following criteria in their definition, each of which is viewed as a continuum and indicative of greater or lesser degrees of community: (1) membership is valued as an end in itself, not merely as a means to other ends; (2) members share commitment to a common purpose and; (3) members have enduring and extensive personal contact with each other.

More recent research has extended the concept of community by exploring the nature and stability of connections between organizations and individuals. In a comparative analysis of private and public schools, Coleman and Hoffer (1987) examine the impact of community on the degree of social integration between families and schools. The researchers assert that the type and strength of community in schools differentially affect the critical social connections which bond families and schools in the joint enterprise of education. This concept of community refers to two types: functional and value. Functional communities are characterized by structural consistency between generations in which social norms and sanctions arise out of the social structure itself, and both reinforce and perpetuate that structure (Coleman and Hoffer, 1987). Functional communities exhibit a high degree of uniformity and cohesion within geographical, social, economic, and ideological boundaries. Value communities describe a collection of people who share similar values about education and childrearing but who are not a functional community; they are strangers from various neighborhoods, backgrounds, and occupations united around an educational organization – their children's school.

The families of neighborhood school students may possess few if any of the constitutive elements of either a functional or value community. While public neighborhood schools a century ago served residential areas that were functional communities, social and technological changes have transformed many of these communities from enclaves of shared values and daily face-to-face talk, to somewhat disparate sets of interests and weak affiliations (Smrekar, 1996).

Workplace schools raise interesting questions regarding the nature and social consequences of a community created under these relatively new and unusual structural agreements between local school districts and employers. Do these hybrid communities promote the uniformity and interconnectedness of functional communities, the shared norms and social trust of value communities, and the "shared space" of traditional residential communities?

School Choice and Community

Many proponents of school choice argue that it will result in greater parent involvement, satisfaction, empowerment, and sense of community (Bryk, Lee & Smith, 1990; Chubb & Moe, 1990; Raywid, 1989). Choice may increase communication between home and school and promote parent commitment to that school, elements which are crucial in developing a stronger sense of community and communal opportunities to learn. As Cookson notes, "Choice enables families . . . to come together in a common effort; school choice rests on social trust and if designed properly, school choice plans can build social trust" (1993, p. 2). Advocates of school choice suggest that choice can integrate families into school communities in ways that may break down the traditional barriers that isolate schools from parents through a supportive, caring climate. Some research has indicated that schools of choice, especially Catholic schools and public magnet schools with a clear, focused mission have higher levels of parent involvement (Bauch & Goldring, 1995; Coleman & Hoffer, 1987). Skeptics, however, contend that school choice may further the fragmentation of communities already splintered through disinvestment, travel to work outside of the area, and forced busing strategies (Moore & Davenport, 1989).

At present, the research on the dynamic intersection of community and choice is primarily speculative and conceptual. Although much empirical research has been conducted on home-school relationships in general, only limited research has focused on home-school relationships in schools of choice. In fact much of the literature on school choice and parental involvement is emerging as two, separate discourses (see Bauch & Goldring, 1995). Research that has been conducted has focused on parent satisfaction with schools of choice, rather than patterns of involvement, communication and commitment (Witte, 1996; Driscoll, 1991). Our recently completed research funded by The Spencer Foundation provides an empirical bridge across the concept of community and the conditions of choice (see Smrekar & Goldring, 1999).

School Choice, Community, and Magnet Schools

One of the primary claims of choice advocates rests with the enhanced social cohesion, commitment, and sense of community engendered in a context of school choice. In earlier work funded by The Spencer Foundation, we explored the conditions in magnet and non-magnet schools and found a somewhat more complicated and confounding story of school community (Smrekar & Goldring, 1999). Certainly, magnet school parents perceive their schools to offer elements that are constitutive of community – a sense of caring and support, a perception of shared values and goals; these parents tend to be a bit more involved in school events than are non-magnet parents. But these findings are muddled by the rather low levels of involvement overall (confirmed by our survey and interview data among both parents and teachers). Moreover, the accounts recorded by the in-depth interviews with parents and teachers indicate little familiarity or face-to-face social interaction among parents or between teachers and parents. The geographical dispersion among magnet school families creates certain social distances that cannot be overcome by the "remade" or reconstituted community that may be miles and miles away from magnet families' neighborhoods. In the absence of any special outreach initiative or creative community-building strategy, this lack of "shared space" in a context of "shared meanings" (Driscoll, 1995, p. 219), diminishes the nature and quality of parental involvement and school community in magnet schools, and renders a far less positive assessment of family-school interactions than some policy makers and choice advocates would predict.

Other concerns relevant to workplace schools emerge from our study of magnet schools. These issues rest with the condition of neighborhood communities that are left behind in the wake of magnet schools. Many magnet parents and teachers in our study lamented the disconnections among neighbors – relationships that in the past were characterized by familiarity and interdependence.

A parent in Cincinnati concluded that in the aftermath of magnet schools, "our communities aren't communities anymore." This observation underscores the concerns outlined by some critics of workplace schools and raises new imperatives for field research on the workplace schools movement.

School Choice, Community, and Workplace Schools

Proponents claim that workplace schools promote the conditions and processes which lead to basic elements of community – sustained commitment, strong communication, and a sense of shared values. As a consequence, these schools of choice can serve a critical function in enhancing educational opportunities and experiences of students and their families (The Education Industry Report, 1997).

On the other hand, workplace schools have been criticized as a strategy that threatens to undermine the foundation of neighborhood public schools. Some argue that workplace schools intensify racial and socio-economic stratification as well as further weaken and fragment the basis for social cohesion and community in neighborhood schools (des Lauriers Cieri, 1997). This work suggests that workplace schools are elitist, exclusionary, and designed as a recruiting tool for only the most highly skilled, sought-after employees. Critics include Theodore Sizer, professor of education at Brown University and the director of the Coalition for Essential Schools project. Sizer argues that controlled enrollment in workplace schools provoke critical questions of access and equity in public education: Who is well served by workplace schools? Who is not?

Questions of access and equity are underscored by the growing efforts to modify enrollment policies under existing state charter school laws to include parents' worksite as the "residence of record" for defining attendance areas. The Medical Center Charter School in Houston illustrates this trend. Similar to other charter schools operating under the open enrollment provisions of the Texas charter school law, the Medical Center school defines its attendance zone by local zip code. In this case, however, the school is open to students whose parents live *or work* in the school's attendance area, a zip code that includes (and is consumed by) the sprawling medical center complex. Few families actually live in the zip code attendance area, so most Medical Center Charter School students are children of medical center employees, many of whom live several school districts away. While some may suggest that these unusual enrollment arrangements make sense in a context in which parents commute long distances to work and are subject to unpredictable and demanding work schedules, the Medical Center Charter School raises important questions about public investments in the private work lives of employees and their families. Thus, at the core of this proposed research rests a fundamental question: Do these school choice initiatives serve a public purpose when "the communities around which people build their lives is their place of work, and not the scattered neighborhoods where they live" (Schnaiberg, March 25, 1998, p. 8)?

Toward a New Theory of Social Integration

Workplace schools embrace John Dewey's (1900) notion of community schools described a century ago; they parallel the school-community partnership model evidenced in school-linked social services (Dryfoos, 1994), and connect with the ideas of community development anchored to Enterprise Zones (Schorr, 1997). They share some of the valued principles of private-public compacts in the movement toward "contracted-out" service arrangements in public schools (Hill, Pierce & Guthrie, 1997). But perhaps most fundamental to this research project is the view that workplace schools reflect the *interdependence* of school, family, and community – a concept that John Goodlad (1987) has labeled "the new ecology of schooling." From the vantage point of employers and educators, this interdependence signals a growing recognition that the separation of work life from family life is undermined in an era of competitive labor markets, dual-career families, and the social press to materially succeed in both work and family (including schooling) endeavors. Traditionally, family-driven work issues (leave, changed meeting times, job sharing) have been accommodated on an individual basis negotiated by corporate Human Resource departments (Bailyn, 1997). This practice is slowly giving way in many organizations to a belief that the collective interests of employees are better addressed through strategic planning that involves decisions beyond the institutionalized ones owned by HR departments. These policies affirm the interdependence of work and family for both organizational and employee benefit. In the 21st century, the family has been made visible in the social milieu of work life.

The compelling interest in this proposed research project rests with the degree to which the workplace school movement signifies a new theory of social *integration* among social institutions. What are the implications of this shift for issues of authority and control? Whose values and what priorities are preferenced in these new social covenants? From a theoretical perspective, this study provides an opportunity to link the emphasis on school and community integration to broader social theory.

SPECIFIC AIMS

The study is conceptually framed by contemporary notions of community and the ecology of schooling that are grounded in social structures and social relations in schools (Bronfenbrenner, Moen & Garbarino, 1984; Bryk & Driscoll, 1988; Coleman, 1987, 1988; Coleman & Hoffer, 1987; Driscoll, 1995; Goodlad, 1987; Merz & Furman, 1997; Smrekar, 1996). Central research questions include:

Choice, Access and Equity
- Who chooses workplace schools and why?
- What is the impact of these choices on the socio-economic diversity of workplace schools?
- What are the implications of these public schools of choice on nearby neighborhood schools in terms of resources, personnel, enrollments, and climate?

Parental Involvement and Community
- What is the impact of workplace schools on the nature and quality of family-school interactions and parental involvement?
- How are parents' social networks affected by these arrangements?
- What are the implications of these new arrangements for information flows and communication patterns among parents of different socio-cultural backgrounds?
- What are the implications of workplace schools for new concepts of family-school-community "partnerships"?

Social Capital and Social Integration
- Do workplace schools promote social capital?
- How do workplace schools affect the roles, responsibilities, and expectations of teachers and principals?
- How do corporate cultures impact workplace schools?
- What is the relationship between (public) workplace schools and the material resources and human capital associated with their (private) corporate sponsors?

METHODOLOGY

The research questions generated by the conceptual framework are grounded in the research literature on school choice, community, and social capital. These questions shape the methodological strategy adopted for this study.

An analysis of workplace schools across four sites provides a template to consider the relationship between the context and character of workplace schools, their organizational components (Bryk & Driscoll, 1988), and the attendant issues of reconstituting community, building social capital, and promoting diversity in American public schools.

An ethnographic, multiple-case study methodology (Yin, 1989) has been adopted for this study and is well suited to explore the issues prompted by the research questions.[1] The intent rests with providing rich, descriptive analysis of events, interactions, and experiences; in interpreting the value and meaning these represent to family, school, and corporate community members; and in translating this knowledge to a set of policy proposals for educators and policymakers at the state and local level that address: parental involvement in public schools; private-public partnerships in education; and educational equity and excellence in the context of school choice.

The sites[2] selected for the study include: the Midwestern Downtown School (sponsored by The American Equities Group), Medical Center of the South Satellite School (sponsored by the Medical Center of the South), Metro Center Downtown School (sponsored by Home Express Stores), and Western Technological Satellite School (sponsored by Virtual Technologies). These school sites were selected using the following criteria: (1) grade levels (at least three, e.g, kindergarten – second grade); (2) size of the school (at least 100 students); (3) the level of corporate investment (resources other than donated land); (4) the number of years the school has operated (at least two); (5) corporate sector diversity (financial, medical, retail, technology); and (6) socio-economic diversity (high salaried management/professional to lower wage low-skilled). The schools selected for this study meet or exceed these criteria.

INITIAL FINDINGS, EMERGING THEMES

I have completed three site visits (3–4 days each) to each of the four workplace schools. During these visits, I have collected school reports, financial records, and numerous other school documents, observed faculty meetings and informal parent-teacher interactions, and conducted formal interviews with teachers and principals, corporate leaders and parents. Early analysis of these data suggest a set of emerging issues related to: (1) the concept of neighborhood as a zone of production (Driscoll & Kerchner, 1999) – social, economic and educational; (2) the meaning of "public" in an era of expansive partnerships between public schools and private organizations; (3) the paradoxical nature of leadership in schools that embrace both interdependence and autonomy, and; (4) the dilemma of perceived elitism and social capital formation relative to a theory of democratic education. The next section considers these issues as they relate to one workplace school and its corporate partners.

MIDWESTERN DOWNTOWN SCHOOL:
THE SOCIAL CAPITAL FOR EDUCATIONAL
AND ECONOMIC REVITALIZATION

The Midwestern Downtown School emerged in 1993 from a consensus view among business leaders that establishing a nationally recognized, "best practices" elementary school was the best strategy for kick-starting broad-based education reform. The "model" school would embrace a central and well-established principle for producing enhanced educational experiences for children – the involvement of parents. With this goal at the forefront of their reform policy, the Midwestern City Business/Education Alliance advanced with a strategy focused upon making work and education more compatible – a notion that sought changes in the social organization of schools while recognizing the pressing demands of parents' work lives.

A downtown location for the new school offered important opportunities to business organizations to "bring the community back to downtown" and in the process, slow the rapidly escalating business flight from central Midwestern City to the new growth areas in nearby suburban West City. The new school fitted neatly with the planned revitalization already under development that involved new housing and entertainment venues nested within a downtown neighborhood. The Business/Education Alliance banked on the perceived value of convenience factors undergirding a "center for living" concept. A high profile, high quality school would attract parent-employees, anchor businesses, and ultimately better serve clients' needs by promoting educational excellence and training for Midwestern City children.

Reluctant initially to expend resources for a new school, the school district eventually accepted the idea that a school located in a downtown Midwestern City office building could solve effectively the problem of parents disengaged from children's school lives. This idea signaled a way of reconnecting physically and symbolically with mothers and fathers whose frenetic work lives drained any discretionary time that otherwise could be spent for volunteer activities at the "neighborhood" public school. Robust and reliable parent involvement would naturally occur under these conditions of convenience, bringing teachers and parents together in the morning rituals of dropping off children at the skywalk school located within a minute's walk to one's nearby downtown office door. Parents would be welcome any time to stop by and observe class instruction, take their child to lunch, or read to a small group of children. As the principal observed in our recent interview: "We've changed what parent involvement means. We're re-connecting with families, making it easier for parents to stop by for 10 minutes in a day. It's perfect for teachers, employees, and employers."

The idea of creating a new space for a school and new social relations with workers-parents in the commercial neighborhood where they dwell during the daytime hours, found credibility and ownership among key stakeholders, including teachers, administrators, parents, and the executives at three major private employers in downtown Midwestern City: the American Equities Group, Digital TelComm, and Danforth Agricultural Research Corporation. These companies agreed to fund any costs above those of an ordinary neighborhood public school, based upon the firm belief that *public* resources should pay for the infrastructure and core technology of *public* schools – bricks and mortar, books and paper. The private partners would provide resources to change the delivery of instruction and enhance the character and quality of public education; the school district would assume "ordinary" school costs. Soon, however, these boundaries of obligation and authority were blurred in a set of extraordinary agreements between the private business interests and the public school district.

American Equities Group donated building space (the company continues to make the lease payments) on the second floor of a downtown office building that was formerly occupied by a bank. The space was redesigned to accommodate 75 children in a semi-open spatial arrangement of round and rectangular tables, blue plastic chairs, low rise stack-style bookcases, painting easels, Apple Macintosh work stations, and a comfortable meeting/faculty room. Two parents – both professional architects who heard about the school through their social networks at a downtown pre-school – volunteered to make the office space fire code compliant for the school district. American Equities Group also offered to underwrite the cost of all professional development activities for teachers and agreed to cover the per diem costs of additional instructional days for a year-round school calendar (at a cost of approximately $100,000 per year).

Some of the institutionalized practices of the Midwestern City Public School District were deconstructed during the process of school change and reinvention with the business organizations. At one pivotal point in negotiations, school district officials agreed to an unusual plan to reallocate resources. To help defray the costs of smaller class sizes (no more than 16) that are triggered by the need for additional classroom teachers, the Downtown School was allowed to forfeit the full-time ancillary teaching staff normally assigned to elementary schools, including a full-time art teacher, music teacher, and a physical education teacher. The school saved other monies by reducing the number of weekly visits by a district school counselor and a school nurse to a minimum.

In the first year, 45 children enrolled in the tiny elementary school. The "Downtown School" was highlighted in a local newspaper article in the summer of 1993 but was not advertised by the school district in any way; information regarding the "experiment" spread by word of mouth among the employees of American Equities, Digital TelComm, and Danforth Agricultural Research. One parent interviewed for this study recalled the first year as one "based upon a blind faith in a school that felt like a big family and like a small business."

Site Description

This past fall, The Midwestern Downtown School enrolled 160 kids from ages 5–11 years-old in a non-graded, year-round instruction program directed by a faculty comprised of 10 full-time teachers. The Downtown School accepts students using a first-come, first-serve admissions system. A sibling priority is applied. The school population is about 25% low-income (eligible for free lunch), although the number is closer to 40% among the families of younger children in the school. Approximately 28% of the children are minority (close to the district average). For the past couple of years, the waiting list for the entering class of five-year olds has numbered over 100 for 32 available slots.

No transportation is provided for students. Over 90% of the parents are employed and most of these work in downtown Midwestern City in occupations that range from attorney and financial analyst, to waitress and data entry clerk.[3]

The school has grown from the original 5000 square feet of office space on the second floor of 501 Hill Street, to include two additional nearby office suites. Each site is separated by less than a three minute walk through the downtown Midwestern City skywalk.

The original Hill Street site is currently occupied by 80 students ages 5–9, five teachers, the school secretary, and the Executive Director (principal). The closest "neighbor" to the school on Hill Street is the Eagle Pantry, an all-purpose general convenience store located a mere 15 feet across the skywalk from the school's front door. In addition to an assortment of fresh fruit, 50-cent steamed hot dogs, the Pantry stocks customary liquor store items, including spirits, beer and wine, assorted tobacco products, lottery tickets, and *Hustler* magazine. Other neighbors, far less colorful and controversial, include a shoeshine stand, a Mexican restaurant, The Bank of America, and the Insurance Exchange.

Three teachers and 48 children (ages 8 through 11) occupy the Capital Square site – 6,000 square feet of renovated space in a corner suite on the first floor of the Capital Square office building, next door to Kinko's Copies. Danforth Agricultural Research Corporation, one of the original three key corporate sponsors of the Downtown School, occupies the top floor of the building and pays the total cost of the lease for the school site. Office building neighbors include a real estate management firm, a snack shop, and a travel agency. The Marriott Hotel and the Midwestern City Convention Center are located on the same street skyway.

Smaller classroom space is leased from a child care center located atop a nearby office building/parking garage; two teachers and 32 children ages 5 through 7 utilize this site. A local University Head Start program is also housed there. Down the hallway a few yards away and sharing the elevator to the 5th floor in this corner of the "tower" building, are the Betty Dixon Manor Apartments for senior citizens – providing an inter-generational reference for the children at the Downtown School.

Today, private investment in the Downtown School is spearheaded by four organizations, three of whom are the "founding members" of the Business/ Education Alliance. The American Equities Group donates approximately $130,000 each year, including lease costs and teacher salary expenditures. Danforth Agricultural Research Corporation contributes approximately $85,000 each year, including the lease costs for the Capital Square site. Digital TelComm completes the original group of three with an annual contribution of $15,000, reduced from an original $25,000 annual donation. This contribution is earmarked for computers and other technology equipment. Universal Insurance Corporation, a new resident of downtown Midwestern City, recently donated $40,000 for a new playground surface and equipment, and committed to a $240,000 gift over the next three years.

Connecting Learning with Life in the City: Social Integration of Work, Family, and School

Against a backdrop of concrete and steel, the persistent threat of crime, and the annoying constraints imposed by scarce commercial space, The Downtown School embraces its urban geography in defiance of these limitations. Instead of defeat and disillusionment, the teachers and students incorporate the city neighborhood as their school without walls. This "sense of place" (Driscoll & Kerchner, 1999) pays rich dividends to teachers, parents, and community members who nurture the notion that the school's location is much more than an address.

For teachers, the freedom to be innovative in the production of knowledge is a manifest element of the Downtown School ethos. This translates into an integrated, seamless connection to the physical and cultural landscape enveloping the school. For a study of physics, employing the project method,[4] teachers "borrowed" the skating rink at the Veterans' Auditorium two blocks away. Lessons were followed by demonstrations of ice-making and melting, and of course, ice skating. Nearby, the highly regarded Midwestern City Art Museum, where a handful of parents of Downtown School children work, provides a free and readily accessible material lesson in art history, architecture, and sculpture. The City Library is the school's library – and the collection is impressive. An invitation issued by the Jefferson Hotel (an architecturally distinctive landmark listed on the Historical Registry) last October involved students in a pumpkin judging contest. The four-minute walk down Hill Street gave teachers the opportunity to pepper students with questions regarding their earlier lessons on "pumpkin math" (geometry of pumpkin cutting), appropriate standards of aesthetic quality, and other essential measurement criteria. The students' artwork is regularly displayed on the walls of the skywalk, the corridors of American Equities Group, and in the windows of the Convention Center. Each year, students from the Downtown School make the five-minute skywalk trip to the Bank of America where bank officials (some of whom are parents of Downtown students) relate the concepts of mathematics, currency, and investment strategies to the children. For teachers, these events are considered "field work" (not an isolated "trip") and are wrapped around the curriculum in fundamental ways that are designed to capture the downtown neighborhood as canvas for students' learning and expression.

The downside to the urban geography is obvious. The threat of crime is a constant, nagging concern. In fact, the first parent committee formed in 1993 when the school opened was the "Security Committee," whose members moved quickly to install a secure entrance. Visitors "buzz in" after a visual identification is made by the secretary behind the glass windows. All three sites now employ this locked door security system and the police are vigilant about notifying teachers of reports of registered sex offenders in the area. The Eagle Pantry is particularly worrisome for the Downtown School because the liquor store is a popular spot for drunk, disorderly drifters in the Downtown area. Recently, when the state lottery reached $140 million, by noon the line of customers at the Pantry laced the length of the skyway.

Until recently, there was very little space for the children to play in this downtown neighborhood. The rooftop at the Tower site includes an open air space with sturdy plastic play equipment but the cold, brisk Midwestern winter and early spring weather often precludes this option for students. So during the cold months, physical education moves over to the nearby Convention Center rooms. In milder weather, the children can play tag and jump rope in the plaza tucked beside the north end of the Capital Square building. Scattered among the dozen or so workers who elect to sit outside in the plaza and enjoy a lunch break, the children play under the close scrutiny of parent volunteers and teachers. This year, however, renovation work on an empty downtown lot now known as "the children's garden," was completed. This grassy space with new play equipment is located on a busy corner just across the street from the Capital Square site. The struggle to bring this dedicated space to the children downtown illustrates the essentials of leadership in the complex, interdependent environment of workplace schools where the cultural chasms between schools, municipal government, and corporate life have been bridged.

Leadership, Autonomy, and Interdependence

For as long as anyone can remember, the empty dirt lot was encircled by a rusty chain link fence that was leaning in several places. Owned by the nearby Jefferson Hotel, the space was slated for a new employee parking lot. The principal recognized that this need could be addressed by utilizing available spaces in a nearby parking garage owned by the City, so she contacted both parties and began a series of conversations. The strategy involved getting the city to agree to a precedent-setting arrangement with a private entity (the hotel), whose owner in turn would agree to "give" the open lot to the school. She earned the interest and sympathy of Midwestern City officials by informing them (none had never heard of the Downtown School) that the school district did not own any land downtown; consequently, the kids had no place to play. Recognizing that the City's parking garage offered little appeal for a playground and represented a danger to schoolchildren, the City moved forward to consider the unusual arrangement.

In an illustrative display of entrepreneurial leadership documented in empirical studies of inter-institutional collaboration (see Smylie, Crowson, Chou & Levin, 1996), the principal established the value of interdependence among the stakeholders. Key strategies involved identifying players' mutual goals; capturing resources; circumventing channels of control and authority (e.g. school district and hotel attorneys); and building a coalition of support to redirect resources (land, parking spaces). The quid pro quo analysis addressed the values (efficiency), priorities (cost effectiveness), and legal structures (accountability) of both city government and a privately held company. Establishing cultural congruence and movement toward collaboration with the school's environment demanded that the principal demonstrate an understanding of separate interests and reward structures among the "active institutional players" (Crowson & Boyd, 1996, p. 157). The principal's actions demonstrated that the school is far less bureaucratic than most with clear (rather than ambiguous), focused (rather than diffuse), humanistic and utilitarian policy goals (Kahne, 1996).

In this case, an agreement recognizing that both social goals (a "common good") and structural ones could be achieved simultaneously, was cemented just in time. Trucks on their way to pave the new parking lot for hotel employees were halted by the hotel owner 30 minutes before scheduled delivery. Today, parking spaces for the employees from the Jefferson Hotel are reserved at the City's parking garage. In exchange for the hotel's payment on these spaces, the City covers the lease on the corner lot. The Downtown School has a safe, grassy playground for students and a new garden that they share with the elderly residents of the Betty Dixon Manor Apartments.

The Product of their Work: Reconstituted Community

The workplace school and the downtown community in Midwestern City have established a set of social relations that provide the constitutive elements of social capital – shared values that lead to trusting relationships and purposeful actions (Coleman, 1987). This fusion of family, work, neighborhood, and school echoes the instructive observations of Mary Driscoll and Charles Kerchner (1999) in their call to "restore a sense of place to educational administration" (p. 394). The authors point to school-family partnerships (see Epstein, 1994), site-based councils, and coordinated services as illustrative examples of educational practices that build the scaffolding for social capital. Perhaps workplace schools signal one of the most robust designs that focuses on the processes of social capital development through community/economic revitalization.

The early findings from this study suggest that principals in workplace schools understand that permeable boundaries help nurture strong ties between people and neighborhood institutions, between schools and the workplace. As magnets for social activity, educational production, and neighborhood stability, the workplace school embodies this ideal. When the workplace and the neighborhood are fused, as in the Midwestern City Downtown School, the "community" is reconstituted, and the concepts of functional and value communities (Coleman & Hoffer, 1987) merge into something new. The hybrid workplace school community creates new structural agreements between schools and employers/employees in terms of more fluid communication and greater financial and cultural interdependence between schools and workplaces. This social scaffolding promotes the shared expectations that in turn, form the uniformity and interconnectedness of functional communities found generations ago in mill towns (e.g. see the Cannon Company mill town in Kannapolis, North Carolina).

When the reality of the new "culture called to work" moves to foreground in American life, the "shared space and shared meaning" (Driscoll, 1995) of traditional residential communities shifts to a new address. Consonant norms and shared trust are evidenced in the relations between parents and teachers in workplace schools and are made manifest in the exercise of school choice.

Without district-provided transportation and no attempt to advertise, parents must make special efforts to seek out and enroll their children in workplace schools. This affirmative decision among people who share similar values about education and childrearing, but who are strangers from various neighborhoods (in fact, are often strangers in their workplace), creates the conditions for value communities in the context of workplace schools.

CONCLUSION

The business community in Midwestern City framed their mission around a mutually agreed-upon goal to shatter the "adopt a school" mold by providing the resources to fundamentally alter and enhance instruction and learning in public education. The imperatives expressed by corporate interests moved well beyond add-ons and extras for classrooms (e.g. computers and library books), and signaled a new interdependence between workplaces and schools. Discontinuities between work lives and the organization of schools were particularly vulnerable to tough scrutiny by private business interests. Clearly, the Downtown School reflects new social trends driven by a set of elevated workplace priorities, illustrating the value of locating a school close to where parents work rather than where they live.

The Midwestern City Business/Education Alliance has produced a partnership that underscores the economic value of social bonds across families, work, and schools. Looking forward, this study will address a set of questions triggered by the Midwestern City workplace school that are relevant in the workplace schools sponsored by Virtual Technologies, Home Express Stores, and Medical Center of the South. How far can the business community move into the discreet sphere of "public" schooling without violating cultural norms and legal sensibilities? Is there a public interest in blurring the boundaries and moving toward full social integration of work, family, school and community? Do workplace schools serve a public purpose or a private one? How do we justify a new school for singular interests (and the perception of privileging particular groups) in a context of scarce resources for neighborhood schools?

NOTES

1. Semi-structured interviews have been completed with the principal and teachers at each of the four school sites, as well as corporate human resource specialists and management personnel involved in the design and implementation of each workplace school. Central office school personnel (e.g. the superintendent) will be interviewed in each of the school districts in which the workplace school is located in order to explore the

"spill over effect" of workplace schools on existing public neighborhood schools (in terms of community support and corporate involvement).

Interviews are underway with parents from each of the four schools, with the goal of interviewing 12–14 sets of parents from each school. Parents will be selected randomly from a stratified sample across social class/occupation groups. School records and parent data cards provide demographic information that can be used to select a sample of parents consistent with the socio-economic and ethnic composition of the total population of school families.

All interviews are audio-taped, with participants' permission, and transcribed verbatim. In addition to interviews, an array of school and corporate documents (including brochures, enrollment applications, letters, newsletters, employee/parent handbooks, and school meeting minutes) have been collected and analyzed. Formal and informal inter-actions involving families and school officials (parent-teacher conferences, workshops, meetings, fundraising events, science fairs, etc.) will be observed over the course of the 20 month data collection period. To ensure anonymity, pseudonyms will be used for individual participants in the study.

Interview transcripts and document analyses will be coded and summarized according to general descriptive categories using the constant comparative method (LeCompte & Preissle, 1993). Converging pieces of information from interview transcripts, field notes, and document analyses will be arranged according to broad themes and categories. Pattern coding (Fetterman, 1989; Miles & Huberman, 1984; Yin, 1989) will be used to discern patterns of thought, action, and behavior among subjects/respondents and schools.

2. School names and the names of their corporate sponsors have been changed to pseudonyms.

3. Approximately 60,000 people work in downtown Midwestern City. According to the principal, many families at the Downtown School are financially stretched because "they can't get the kinds of jobs they would like. It's usually females who cannot find a good full-time job . . . So they work two or three part-time jobs with no benefits. Their life, their schedule, is horrendous."

4. The Midwestern Downtown School has been the focus of local and national media attention for its integrated curriculum/problem-solving approach, small class size (16), multi-age classrooms, experience-based, active learning pedagogy, portfolio system of assessment, enhanced parent involvement, and extended school year. In September, 1999, the Downtown School was recognized by a national magazine for all-around excellence in serving the needs of working parents.

REFERENCES

Bailyn, L. (1997). The impact of corporate culture on work-family integration. In: S. Parasuraman & J. Greenhaus (Eds), *Integrating Work and Family: Challenges and Choices for a Changing World* (pp. 209–219). CN: Quorum.

Bauch, P., & Goldring, E. (1995). Parent involvement and school responsiveness: Facilitating the home-school connection in schools of choice. *Educational Evaluation and Policy Analysis, 17*(1), pp. 1–21.

Bellah, R., Madsen, R., Sullivan, W., Swindler, A., & Tipton, S. (1985). *Habits of the heart*. Berkeley, CA: University of California Press.

Bourdieu, P. (1977). Cultural reproduction and social reproduction. In: J. Karabel & A. Halsey (Eds), *Power and Ideology in Education*. New York: Oxford University Press.

Bronfenbrenner, U., Moen, P., & Garbarino, J. (1984). Child, family, and community. In: R. Parke (Ed.), *Review of Child Development and Research*, Vol. 7. Chicago: University of Chicago Press.

Bryk, A., & Driscoll, M. (1988). *The high school as community: Contextual influences and consequences for students and teachers*. Madison: National Center Effective Secondary Schools.

Bryk, A., Lee, V., & Smith, J. (1990). High school organization and its effects on teachers and students. In: W. Clune & J. Witte (Eds), *Choice and Control in American Education*, Vol 1: *The Theory of Choice and Control in American Education* (pp. 135–226). Bristol, PA: Falmer Press.

Chubb, J., & Moe, T. (1990). *Politics markets and America's schools*. Washington, DC: The Brookings Institution.

Coleman, J. (1987). Families and schools. *Educational Researcher, 16*(6), 32–38.

Coleman, J. S., & Hoffer, T. (1987). *Public and private high schools: The impact of communities*. New York: Basic Books.

Cookson, P. (1993). *School choice and the creation of community*. Paper presented at the conference, Theory and Practice in School Autonomy and Choice: Bringing the Community and the School Back In. Tel Aviv University, Israel.

Cowans, D. (1997, June 23). Onsite schools keep parents close. *Business Insurance, 3*, 6, 9.

Crowson, R., & Boyd, W. (1996). Structure and strategies: Toward an understanding of alternative models for coordinated children's services. In: J. Cibulka & W. Kritek (Eds), *Coordination Among Schools, Families, and Communities: Prospects for Educational Reform* (pp. 137–169). Albany, NY: State University of New York Press.

Dewey, J. (1900). *The school and society*. Chicago: University of Chicago Press.

des Lauriers Cieri, C. (1997, January 6). Off to work? Don't forget to bring the kids. *The Christian Science Monitor, 1*, 12.

Driscoll, M. (1991). *Schools of choice in the public sector: Student and parent clientele*. Paper presented at the annual meeting of the American Educational Research Association, Chicago.

Driscoll, M. (1995). Thinking like a fish: The implications of the image of school community for connections between parents and schools. In: P. Cookson, Jr. & B. Schneider (Eds), *Transforming Schools*. New York: Garland.

Driscoll, M., & Kerchner, C. (1999). The implications of social capital for schools, communities, and cities: Educational administration as if a sense of place mattered. In: J. Murphy & K. Seashore Louis (Eds), *Handbook of Research on Educational Administration* (pp. 385–404). San Francisco: Jossey-Bass.

Dryfoos, J. (1994). *Full-service schools*. San Francisco: Jossey-Bass.

Elshtain, J. (1995). *Democracy on trial*. New York: Basic Books.

Fetterman, D. M. (1989). *Ethnography step by step*. Newbury Park, CA: Sage.

Galinsky, E., Bond, J., & Friedman, D. (1993). *The changing workforce: Highlights of the national study*. New York: Families and Work Institute.

Gallup, G., & Newport, F. (1990). Time at a premium for many Americans. *Gallup Poll Monthly*, November, 43–56.

Goodlad, J. (1987). *The ecology of school renewal. Eighty-ninth Yearbook of the National Society for the Study of Education*. Chicago: University of Chicago Press.

Hill, P., Pierce, L., & Guthrie, J. (1997). *Reinventing public education*. Chicago: University of Chicago Press.

Hochschild, A. R. (1997). *The time bind*. New York: Metropolitan Books.

International Labor Organization (1999). *Key indicators of the labor market 1999*. Geneva, Switzerland: ILO.

Kahne, J. (1996). *Reframing educational policy*. New York: Teachers College Press.

Lareau, A. (1989). *Home advantage*. New York: Falmer Press.

Lasch, C. (1995). *The revolt of the elites and the betrayal of democracy*. New York: W.W. Norton.

LeCompte, M., & Priessle, J. (1993). *Ethnography and qualitative design in educational research*. San Diego, CA: Academic Press.

Merz, C., & Furman, G. (1997). *Community and schools*. New York: Teachers College Press.

Miles, M., & Huberman, A. (1984). *Qualitative data analysis*. Beverly Hills, CA: Sage.

Miller, K. (1997, November 9). *Workplace schools continue to grow*. Associated Press Online.

Moore, D., & Davenport, S. (1989). *The new improved sorting machine*. Chicago: Designs for Change.

Munk, N. (1996, September 9). *Good schools, good business*. Forbes, 144–148.

Murray, B. (1998, January 5). How to give working parents more time with their kids. *U.S. News and World Report*, 88–90.

Newmann, F., & Oliver, D. (Winter 1968). Education and community. *Harvard Educational Review*, 37, 61–106.

Orfield, G., Bachmeier, M., James, D., & Eitle, T. (1997, April). *Deepening segregation in American public schools*. Cambridge, MA: Harvard Project on School Desegregation.

Putnam, R. D. (1995). Bowling alone: America's declining social capital. *Journal of Democracy*, 6(1), 65–78.

Raywid, M. (1989). The mounting case for schools of choice. In: J. Nathan (Ed.), *Public Schools by Choice: Expanding Opportunities for Parents, Students and Educators*. St. Paul, MN: Institute for Learning and Teaching.

Scherer, J. (1972). *Contemporary community: Sociological illusion or reality*. London: Tavistock.

Schnaiberg, L. (1998, March 25). Worksite charter schools take the edge off commuting. *Education Week*, XVII(28), 8–9.

Schor, J. (1992). *The overworked American: The unexpected decline of leisure*. New York: Basic Books.

Schorr, L. (1997). *Common purpose*. New York: Anchor.

Schultz, T. (1963). *The economic value of education*. New York: Columbia University Press.

Smrekar, C., & Goldring, E. (in press). *Magnet schools in urban districts: What is our choice?* New York: Teachers College Press.

Smrekar, C. (1996). *The impact of school choice and community: In the interest of families and schools*. Albany, NY: State University of New York Press.

Smylie, M., Crowson, R., Hare, V., & Levin, R. (1996). The principal and community-school connections in Chicago's radical reform. In: J. Cibulka & W. Kritek (Eds), *Coordination Among Schools, Families, and Communities: Prospects for Educational Reform* (pp. 171–195). Albany, NY: State University of New York Press.

Steinberg, L. (1989). Communities of families and education. In: W. Weston (Ed.), *Education and the American Family*. New York: New York University Press.

The Education Industry Report (1997, April). *Schools at work: Providing a win-win situation*. Vol. 5(4).

U.S. Bureau of Labor Statistics (1989). *Handbook of labor statistics*. Washington, D.C.: U.S. Bureau of Labor Statistics.

Witte, J. F. (1996). Who benefits from the Milwaukee choice program?. In: B. Fuller, R. Elmore, & G. Orfield (Eds), *Who Chooses? Who Loses? Culture, Institutions, and the Unequal Effects of School Choice* (pp. 25–49). New York: Teachers College Press.

Yin, R. (1989). *Case study research*. Newbury Park, CA: Sage Publications.

CIVIC CAPACITY AND SCHOOL PRINCIPALS: THE MISSING LINKS FOR COMMUNITY DEVELOPMENT

Ellen B. Goldring and Charles Hausman

INTRODUCTION

Recently, we had the opportunity to meet with a new principal of an urban elementary school located in a run-down industrial part of town, serving two public housing projects. She was appointed as the principal of an *enhanced option school*, a school slated to provide services and educational options to its students, parents, and the community at large. She expressed examples of the need for community services:

> We need health services. We cannot have a situation where children are not enrolled because they don't have their shots. Well, the goal is if you can do that here, they can go to class. Why wait? So, it is another extension of a satellite, just to have the nurses twice a week, would be a plus. The common thing we have, for example, we deal with head lice. I have to check. They have to go down to the health department, have someone say, you are nit free to come back to class. We need parents who can bring their kids to a school nurse so they can very quickly go to class. This would really help parents because for a lot of parents transportation is a dilemma. You can't get there, you have to go downtown, and then transfer a bus . . . So children just don't come to school.

What does it take to mount a program where health services are located at the school? What does it take for a school to offer hot breakfast to children on holidays and during summers, knowing that the school is the only institution

Community Development and School Reform, Volume 5, pages 193–209.
Copyright © 2001 by Elsevier Science Ltd.
All rights of reproduction in any form reserved.
ISBN: 0-7623-0779-X

in the local community providing nutritious meals to these children? What does it take to revitalize the surrounding neighborhoods of schools so entrepreneurial businesses will locate in isolated buildings? This paper argues that civic capacity – "the ability to build and maintain an effective alliance among institutional representatives in the public, private, and independent sectors to work toward a common community goal" (Henig, 1994, p. 220) is key to addressing the needs of urban schools. Furthermore, urban school principals must be the key champions and stewards of developing communal civic capacity to support widespread community building efforts. Principals can develop civic capacity by forming partnerships to garner additional resources from the business community and by serving as central members of key stakeholder groups. Principals must work closely with community and social agencies that assist students and their families. Broad-based coalitions must be formed with civic responsibility to rejuvenate neighborhoods where schools are located. Education takes place within the context of communities, not just in a school building.

This chapter reviews the notion of civic capacity and asks questions about its centrality to principals' roles in urban schools. We examine the extent to which elementary school principals in two urban school districts serve as builders of civic capacity. The paper concludes with a discussion about civic capacity and the urban school principals' role in school reform. We argue that civic capacity, the will, capability and drive of communal leaders, is absolutely paramount for schools to become embedded in their local communities. The school principal must play a key role in the development and support of this sense of civic capacity.

CIVIC CAPACITY AND COMMUNITY DEVELOPMENT

Educators have long argued that schools alone often lack the capacity to address the multiple social problems facing students in urban schools, especially as pivotal indicators of social well-being continue to decline. Students arrive at school each day with a myriad of sociological and psychological challenges. Forty percent of children under the age of 18 in America live in poverty (Lugalia, 1998). Dryfoos (1994) reported that one in four children "do it all – use drugs, have early unprotected intercourse, are truant and fall far behind in school" (p. 3). These children are not likely to have access to adequate family supports, health care, and social networks that are crucial for success in school. As a result of these social changes, schools have been relegated several new roles – social worker, health-care provider, character developer, and more. "Schools have become the repositories for societal responsibilities. Kids bring problems into the classroom that society has failed to address on the outside. Then teachers are told to be mother, confessor or educator" (Chavez, 1999, WWW).

Reformers continue to claim schools should be placed at the focal point of entire communities to increase the potential for schools to meet these divergent needs of students, especially those most at risk for failure. Relying on notions about the ecological perspective of schooling (Bronfenbrenner, 1979), educators emphasize the importance of social linkages between children and adults across different environments and structures. Support systems that reach multiple contexts for children often act as "immunizing" factors against adversity (Wang, 1997; McLaughlin, 1994).

Educational reforms have relied on various efforts to increase community-school linkages: community control, parental involvement, parental choice, and coordinated services, to name a few (Beck & Foster, 1999). The last decade reported movement towards coordinated services in schools or full-service schools (Dryfoos, 1994). These schools offer an array of services for children *in schools* from health clinics, to employment counseling and parent education. These initiatives have been implemented across the country with varying degrees of success. Issues of turf, funding, politics, confidentiality, and trust all caused challenges to the successful implementation of social service delivery in schools (Smrekar, 1996).

As researchers were giving mixed marks to full-service schools (Crowson & Boyd, 1993), new federal initiatives, such as empowerment zones and enterprise communities, re-focused attention to broad-based community initiatives to address the deepening distress in many urban communities (Empowerment, 1995). Community development initiatives often focus on the economic development of urban centers as a first priority for community revitalization. Thus, private foundations, banks, and neighborhood agencies join hand and hand for the *total* re-development of a community, rather than merely placing expanded services in the schools. The premise behind these newer models of school-community partnership, often referred to as community development initiatives, is the assumption that schools are central to an economic revival of central cities. "At its heart, community development constitutes a philosophical change in the way schooling is conceived. Community development changes the core identity of schools form isolated, independent agencies to institutions enmeshed with other community agencies in an interconnected landscape of support for the well-being of students and learners. It beckons schools to consider and respond to learning needs throughout the community, not just to those of children within the school building" (Timpane & Reich, 1997, p. 466).

Beginning to emerge in urban centers are examples of community development initiatives that involve schools (Cumo & Glickman, 1998):

- As part of the Enterprise Community of New Orleans, initiatives at Safe
  Harbor Schools encompass parenting centers, youth outreach to re-engage
  truant and dropouts, and programs to enhance graduation through computer
  labs, job readiness training and multiple supports with community agencies.
- Success by Six is part of the Little Rock, Arkansas enterprise community. It
  is a program supported by 10 key stakeholders and a steering committee
  representing more than 50 individuals and organizations. The program
  supports home visits to coordinate household needs and community services
  and educational programs.
- The enterprise community of New Haven sponsors youth Fair Chance. This
  program supports youth in the city, including teen clubs, trips, sports and job
  training. The program also collaborates with a school for returned dropouts,
  where guidance counselors collaborate with community officials to develop
  internships. A computer assembly class, for example, has resulted in students
  being hired by EC council members to market and sell computers.

In theory, then, community development initiatives embrace a perspective that
suggests that schools and their communities must be linked in multiple avenues,
multiple spaces, and multiple activities.

Lessons learned from years and years of community partnerships suggest that
multiple stakeholders, including educational professionals, politicians, business
leaders, community agencies and parents, are needed to support and sustain
community devolvement initiatives (Stone, 1998). The notion of stakeholder
involvement is central to enable communal leaders to serve as levers for change,
as well as supporters and sustainers of change. For example, in an analysis of
school reform in Chicago and New York, Gittell (1994) found that involvement
from parents and their associations, students and community activists, the media,
business leaders, foundations and other "civic style education interests groups"
was the key for the implementation of community-wide education reform. Gittell
stresses the consistency and strength of stakeholder involvement as essential for
the implementation of widespread school reform. "The differing role of these
stakeholders – their strengths and weaknesses and their ability to form coalitions,
mobilize support and resources – affect the output, both educationally and
politically" (Gittell, 1994, p.138). Stakeholder involvement must become a force
for school reform that occurs within a broader community framework.

Recently the concept of *civic capacity* has been applied to the notion of community development and school reform (Orr, 1996). Civic capacity is defined as

the "degree to which a cross-sector coalition comes together in support of a task of community wide importance" (Stone, 1998, p. 234). The concept refers to the ability to build and maintain effective alliances for collective problem solving (Orr, 1996). Civic capacity emphasizes the collective role of community stakeholders, going beyond the view that any one single institution, such as schools, can address the needs of its constituencies. "In the education arena, the capacity of any set of stakeholders is limited, and so long as various players think and act in terms of a narrow view of their duty, they miss the full scope of the problem they face and the response to which they could contribute" (Stone, 1998, p. 254). Civic capacity is the cornerstone of community building. Community building, defined as "continuous, self-renewing efforts by residents and professionals to engage in collective action aimed at problem solving and enrichment that creates new or strengthened social networks, new capacities for group action and support and new standards and expectations for life in the community" (Walsh, 1997, p. 5), can thrive when key professional and lay leaders embrace widespread civic capacity in place of narrow institutional interests.

Stone (1998) suggests that civic capacity needs "renewed appreciation" as a concept applied to the community development discourse. Community development will not happen on its own; people from various agencies, institutions and groups do not naturally come together to solve mutual problems and develop initiatives. Therefore, civic capacity denotes the motivation, community-mindedness, and activities necessary to address community-wide problems that are central to schools. Stone suggests that civic capacity goes far beyond notions of social capital. Social capital relies on the development of reciprocity and trust. Stone claims that social capital does not transfer easily from one context to another, or from one situation to another. "Social capital also rests, in significant part, on a basis of *shared loyalty and duty*. Reciprocity and trust are thus circumscribed; they may apply within some circles but not in others" (Stone, 1998a, p. 268). Community development must engage individuals in an intergroup context, not in an interpersonal context; hence the idea of civic capacity.

Social capital, or interpersonal trust, can be a starting point for developing civic capacity, whereas civic capacity involves the need for communal groups to put aside competitive impulses and focus on community-wide alliances. Brown (1998), in fact, illustrates that cooperative relationships and mutual problem solving between institutions and organizations can actually contribute to the development of social capital. "If social capital grows out of experiences of successful cooperation across differences in sector and power, agencies that can successfully bridge those gaps in order to promote cooperation can play an extraordinary important role" (Brown, 1998, p. 240).

Civic capacity tends to focus community building initiatives on five general areas: engaging government systems, building local institutions, investing in outreach and organizing, involving the corporate sector, and developing new structures (Walsh, 1997). All of these types of initiatives can impact and build capacity for and in schools. Engaging government systems requires that schools access and command the resources and supports that are available. For example, the Department of Education for the State of Illinois recently hired a very high-prestigious lobbying firm in Washington, DC to ensure that the state is receiving all possible federal funds. Similarly, local schools that spend time networking with existing government agencies are able to bring services to bear in their institutions. A principal in a local urban school recounted how she goes to community agency meetings in order to network and become known to other professionals in the city to reap the benefits from other government agencies:

> I make initial contact because I'm at a meeting and I happen to meet someone. It is, voila, there it is. I was invited to be part of a group and as a result of that presentation, listening to what they are talking about, I can begin to be connected and express our school's needs. Here we are, here is what we need, and can we get this?

Community development requires mutual support from local institutions that often entails sharing resources to ensure the continued vitality of institutions that can provide services with schools. A local YMCA, for example, running after care programs and youth outreach services are dependent upon the schools to partner, emphasize and integrate programs. Partnerships can include sharing space, advertising and helping parents register for the programs, tackling transportation issues, and co-developing curricula and activities for the children.

Investing in outreach and organizing is essential for community development work. Network building is absolutely necessary to cultivate community partnerships both with the private and public sectors in the community. Business partnerships are no longer limited to tutors coming in to the schools during lunchtime. Real partnerships, with mutual commitment, are needed.

All of these types of community development activities depend on a sense of civic capacity from school as well as community members. Civic capacity "entails not simply bringing a coalition together around the issue of educational improvement but beyond that engaging the members in activities and promoting discourse" (Stone, 1998a, p. 258). Education is framed in "community wide terms".

PRINCIPALS AS BUILDERS OF CIVIC CAPACITY

As decades of research on effective schools and school reform has indicated, it is unlikely that substantial change can occur in the nature of school-community partnerships unless school principals embrace a more community-oriented perspective, that is, unless school principals view the development of civic capacity and community building as part of their roles. Effective schools research emphasizes the "behavior of the school principal is the single most important factor supporting high education quality" and "while schools make a difference in what students learn, principals make a difference in schools" (cited in Bredeson, 1989). However, the notion of civic capacity can create a real dilemma for principals. The dilemma is one of "share of mind". Given the current climate of educational accountability and immediate pressures to "show results", will principals have the "share of mind" to engage in civic capacity? Will principals embrace civic capacity as an avenue for school improvement?

Much has been written about the challenges of collaboration in education (Pounder, 1998). These challenges often focus on incompatible roles, lack of resources, inconsistent cultures, and contradictory structures. Principals often confront similar challenges of control, uncertainty, and accountability as they collaborate with teachers, parents, and community agencies (Crow, 1998). The concept of civic capacity goes beyond notions of partnering or collaborating. It requires a new mental model of schooling. While collaboration and inter-dependence are necessary for school improvement, they are not sufficient. "Educators, service providers, youth development specialists, and others join forces to collaborate because of enlightened self-interest. Self-interest means, for example, that principals invest time, energy, and resources in community and family work because they know that they and their schools cannot be successful without them. They are self-interested but not selfish. They are also enlightened because they choose their involvement strategically with an eye toward building supports for children and schools" (Lawson, forthcoming, p. 12). These supports are highly interwined with the 'health' of the community at large. Since principals are expected "to do it all", this enlightened self-interest and awareness is critical if principals are going to allocate time and energy to building civic capacity.

Principals face unique aspects of the challenge of developing civic capacity in the current climate of high-stakes testing and statewide accountability systems. Most principals are confronted with the ongoing pressures of "showing results" quickly and regularly. Sergiovanni (2000) recounts the typical school year for most principals centering on annual tests. Early in the school year, he suggests, principals help focus schools on shared purposes and values. However, "by late February or early March the focus on tests reaches a frenzy in some schools" (p. 86). This often requires principals to re-focus, and at times abandon their missions, to prepare students for the tests. After the testing in April, it is a mere settling down until the end of school year.

With current accountability pressures, to what extent do principals view community outreach and civic capacity as important in their roles? We relied on data that were collected as part of a larger study of urban elementary schools in Cincinnati and St. Louis. Both of these cities contain large urban school districts with well-established needs in community development and outreach. The Cincinnati school district includes 86 schools: 61 elementary, eight middle/junior high, ten secondary, and seven special schools. The enrollment is approximately 51,000 students, 66% of whom are African-American, 32% white, and 2% other. Sixty-two percent of students in the district are from low-income families, with an average of 379 students from low-income families per school. The city of Cincinnati is 78 square miles in Hamilton County. Since most demographic and economic data are collected at the county level, the following numbers are reported for Hamilton County as whole, noting that while these numbers include Cincinnati, they are more favorable than the data for the city of Cincinnati alone. Twenty-four percent of county residents have less than a high school diploma. The median household income in 1993 was $33,248, and 22.2% of children ages 5–17 lived in poverty. Over 10% of the population does not have health insurance. The crime rate per 1,000 persons is 57.7, and the teen birth rate per 1,000 females, ages 10–19 is 29.8.

The St. Louis City School System (SLPSS) operates 104 schools – 73 elementary, 21 middle, and 10 high schools. The total enrollment is 36,091 – 78% are African-American. SLPSS is a classic example of an urban system confronted with typical, yet devastating, social and economic problems. Since 1950, the city population has decreased by over 50%. Since 1990, 7% of the population relocated out of the city. This exit was predominantly middle class families. This has led to a median family income in the county ($38,500) which is almost double that in the city ($19,458) (Task Force on Desegregation of the St. Louis Public School System, 1995). The net result for the SLPSS is fewer students and students who require more resources to educate. The drop out rate for the district is 37%. Clearly, Cincinnati and St. Louis are typical of large urban centers with schools and children whose needs are far reaching.

We surveyed 46 elementary school principals in these districts and report their sense of involvement with and commitment to civic capacity and community outreach. The Principal Survey was mailed to the head principal of each school in the sample. This package included a return envelope for confidentiality. To further maintain anonymity, completed principal surveys cannot be linked to a specific school. Follow-up letters were mailed to all principals as a reminder and to encourage them to return their questionnaires. Thirty-eight out of 46 principals mailed in completed surveys for an 82.6% response rate. Although this is a small sample size, it is illustrative of the

Table 1. Time Spent on Community Engagement.

Item	None	Very Little	Some	A Great Deal
Working with local community agencies in solving problems	1 (2.7%)	15 (40.5%)	15 (40.5%)	6 (16.2%)
Making arrangements for supportive services for students such as testing and speech therapy	0 (0%)	6 (15.8%)	22 (57.9%)	10 (26.3%)
Cooperating with social agency officials	1 (2.7%)	11 (29.7%)	17 (45.9%)	8 (21.6%)
Obtaining community-based resources to enrich the curriculum	1 (2.7%)	17 (45.9%)	13 (35.1%)	6 (16.2%)
Identifying resources in the community to assist the families of children in the school	1 (2.8%)	11 (30.6%)	15 (41.7%)	9 (25.0%)

challenges principals face in expanding their roles to encompass education beyond the school walls.

Community Engagement

We first explored the extent to which principals spend time engaging with their community in trying to develop civic capacity (see Table 1). Only 16% ($n = 6$) of the principals said they spend a great deal of time working with local community organizations and agencies in solving problems, while 43% ($n = 16$) spend very little or no time with community agencies. Similarly, the same number of principals report spending a great deal of time trying to identify and obtain community resources for their schools, while 49% ($n = 18$) spend very little or no time in this role. In contrast, principals seem more willing to interact with community agencies when their services are directly related to educational services for students and their families. Thus, 26% ($n = 10$) of principals spend a great deal of time arranging for supportive services for students, such as testing and speech therapy, while 60% ($n = 22$) spend some time on negotiating these services that are not provided directly by the school. Similarly, 67% of these principals spend some or a great deal of time identifying resources in the community to assist the families of children in the school. On average, these urban elementary school principals report only modest interaction and networking with social and community agencies. Furthermore, principals of schools serving higher percentages of students from lower socioeconomic backgrounds *do not* spend significantly more time with community agencies than other principals in this sample. The engagement seems to be solely focused on school services rather than embracing a 'community development at large' perspective.

Community Partnerships

Another aspect of civic capacity and community building involves the establishment of partnerships between schools and local business and community organizations (see Table 2). To gauge the extent to which business and local community organizations provide various types of support to local schools, we asked principals to assess the level of support from outside agencies of their schools in various arenas. On average, across six different types of partnerships, these principals reported moderately low levels of support from business and community partnerships. For example, principals reported that local business or community organizations are moderately supportive in terms of donating supplies or equipment and tutoring students. They are even much less supportive in terms of serving as curriculum or program advisors or offering professional development opportunities for teachers or staff.

Table 2. Community Partnerships.

Item	Not at All	Somewhat	Moderately	Very Supportive
Donations of supplies or equipment	3 (7.9%)	17 (44.7%)	9 (23.7%)	9 (23.7%)
Provision of speakers/internships	5 (13.2%)	7 (18.4%)	20 (52.6%)	6 (15.8%)
Serving as curriculum/program advisors	20 (52.6%)	12 (31.6%)	3 (7.9%)	3 (7.9%)
Offering prof. development workshops to teachers/staff	10 (26.3%)	21 (55.3%)	5 (13.2%)	2 (5.3%)
Donation of funds	9 (23.7%)	15 (39.5%)	6 (15.8%)	8 (21.1%)
Tutoring students	5 (13.2%)	15 (39.5%)	8 (21.1%)	10 (26.3%)

Building Civic Capacity

Principals face multiple demands and must prioritize their time. They must decide what is most important for them to take care of, and what can be delegated. Much has been written about the various roles of principals (Rallis & Goldring, 2000). These roles are often thought about in terms of internal roles focusing on student achievement, teacher development and creating a strong internal culture, and external roles, such as interacting with parents, community service agencies and garnering resources. The external roles are congruent with a civic capacity perspective. With this distinction of roles in mind, we asked principals two central questions. First, what roles do principals consider most important? Respondents were asked to prioritize their top three roles as principals. Community building roles included networking with social agencies to better serve changing student populations and interacting with the school's external environment to build support and garner resources. Other roles included in the rankings were instructional leader, providing vision and inspiration for the school, and keeping the school running smoothly. Second, we asked the principals if they had more discretionary time, how would they allocate it?

When asked to rank order the three most important roles from a list of nine, only one principal ranked "networking with social agencies to better serve changing student populations" among their top three, and this role was ranked third most important. Even serving as a caretaker (e.g. maintaining the status quo) was deemed more important amongst our small sample of principals. Similarly, as a group, these principals placed relatively little importance on interacting with the school's external environment to build support and garner resources. Three principals rated this role second most important, while two rated it third (see Table 3). Principals indicate that it is most important to serve as an instructional leader. Thirty-two percent of the principals ($n = 12$) ranked this role as most important, and 34% ($n = 13$) rated this as the second or third most important role. Principals also indicated that it is important to provide a vision for the school as 29% of the principals ranked this as the most important role.

Perhaps it is not surprising that those roles which are well-established in educational circles – cultural leader, facilitative leader, instructional leader, visionary – were reported as most significant in relatively equal numbers. These findings are congruent with the stress on accountability today. These results are a testimonial to the vast number of roles that school administrators must balance. However, the common thread is that these principals view roles enacted *in the school* as vastly more important than roles requiring external interaction.

One possible explanation for the low emphasis on external engagement is time constraints. Clearly, principals are expected "to do it all," but they only have

Table 3. Principal Roles.

Role	Most Important	Second	Third
Serving as the school's instructional leader	12 (31.6%)	6 (15.8%)	7 (18.4%)
Providing vision and inspiration for the school	11 (28.9%)	9 (23.7%)	4 (10.5%)
Providing teachers with the resources and skills they need to best meet the needs of students	7 (18.4%)	9 (23.7%)	8 (21.1%)
Creating a sense of community and school culture conducive to learning	6 (15.8%)	8 (21.1%)	7 (18.4%)
Interacting with the school's external environment to build support and garner resources	0 (0%)	3 (7.9%)	2 (5.3%)
Keeping the school running as long as things are going okay	0 (0%)	0 (0%)	2 (5.3%)
Networking with social agencies to better serve changing student populations	0 (0%)	0 (0%)	1 (2.6%)
Selling the school to students and parents to attract and retain students	0 (0%)	0 (0%)	3 (7.9%)
Making ethical decisions and serving as a positive role model with high values	1 (2.6%)	3 (7.9%)	3 (7.9%)

time, and share of mind, to allocate to what they believe is most important. Principals were asked to report how they would spend time if they could magically find an additional ten hours per week. Developing community outreach and partnerships was one possible option. Only two principals indicated they would allocate this additional time to community outreach and partnerships, while 55.2% reported they would spend it on instructional leadership. Closely related to community development is parental involvement, and only one principal indicated s/he would spend the extra time on interacting with parents. Next to instructional leadership, principals would spend new available time with students, another aspect of the role within the boundaries of the school.

Collectively, these findings suggest that simply finding more time to engage with community leaders may not be the solution. Clearly, the case needs to be made so that principals view the role of builder of civic capacity as important. The infrequency with which these principals work with social/community agencies and businesses and the lack of importance attributed to this role are alarming given the high number of students at risk in this urban sample, the low level of resources characteristic of so many schools today, and the lack of community building activities in the neighborhood at large.

BUILDING CIVIC CAPACITY: IMPLICATIONS FOR PRINCIPALS

The evidence is overwhelming that the crises in education relate not just to school governance but to pathologies that surround the schools. The harsh truth is that, in many communities, the family is a far more imperiled institution than the school, and teachers are being asked to do what parents have not been able to accomplish. Today, the nation's schools are called upon to stop drugs, reduce teenage pregnancy, feed students, improve health, and eliminate gang violence, while still meeting academic standards. And if they fail anywhere along the line we condemn them for not meeting our high-minded expectations.... is it realistic to expect an island of academic excellence in a sea of social crisis? Simply stated, educational excellence relates not just to schools but to communities as well (Carnegie Foundation for the Advancement of Teaching, 1992, p. 76).

As builders of civic capacity, principals must "bridge the connection between the conditions of education and the total conditions of children"(Kirst, McLaughlin & Massell, 1989). Schools, as one of the central intact social institutions in many urban communities, must take a central role in community development. Principals must begin thinking about education in a broader context and allocating additional time to roles enacted external to the school. Building civic capacity is one such critical role. A true commitment to developing civic capacity requires principals to have a much broader view of the place and function of schools in urban communities.

Developing civic capacity surfaces several obstacles and issues for principals. First, the continual, additive nature of the role of principals has been well documented. Many educators and researchers argue that principals already lack sufficient time to perform all of the responsibilities expected of them. Each time new reform efforts, policies and directives emerge, principals are expected to change their foci. Debates still linger in the school administration corridors as to whether principals are leaders or managers, visionaries or implementers, instructional leaders or administrators. Each of these themes carries with them specific implications about how and where principals' enact leadership and with whom and when they spend time with various constituencies and stakeholders.

In many ways, these discussions set up false dichotomies. Principals can think of leadership as an organizational and community-wide phenomenon (Pounder,

Ogawa & Adams, 1995). When viewed from organizational and community-wide units of analyses, leadership is not a zero-sum construct. In other words, the total amount of leadership varies across situations. Moreover, as a broader range of stakeholders share leadership, it is no longer isolated to those holding formal positions, such as principals. Teachers, parents, students, and staff members provide leadership in schools. While viewing leadership as an organization-wide phenomenon highlights multiple sources of leadership internal to the school, leadership may also be viewed as a community-wide construct. Consistent with the notion of civic capacity, leadership can emerge from organizations and individuals external to the school. In fact, civic capacity requires leadership development from all stakeholders.

Viewing leadership in these broader contexts places a premium on the collaborative skills of principals. Collaboration within schools and between schools and other community agencies has proven to be difficult. As schools broaden partnerships with communities, boundaries will be difficult to define and role ambiguity between different providers will likely be enhanced. These should all be viewed as welcome signs of change.

Another challenge of building civic capacity centers on resources. Programs targeted at building civic capacity often are funded by start-up grants. When these grants expire, principals must generate ongoing resources for the programs to be sustainable (Lawson, forthcoming). This will be a tremendous challenge in an era where the benefits of additional resources for education are hotly contested and political pressures for tax reductions are gaining momentum. Large numbers of program coordinators, health service providers, student support professionals, and other agency service providers are needed, and the caseloads of these individuals are already overwhelming, especially in urban areas. This limitation of human resources has constrained some districts and communities from scaling up coordinated service programs after successful pilot studies (Lawson, forthcoming). Principals who are successful builders of civic capacity are tireless advocates for all children-constantly lobbying for the necessary resources to effectively meet the needs of all children.

A recent conference, "*The Business and Education 2000 Conference: Building Strategic Partnerships That Work From The Inside Out*" highlights some of the complexities around resource issues when developing partnerships between corporations and schools. Corporations are moving away from a general adopt a school model, to follow a more strategic approach to target specific, long-range programs. For example, one such corporation is center.ing efforts on integrating technology in a set of schools in a district. The corporation is not simply donating computers but is focusing on teacher training and students skill development for the future workforce. School administrators, however, often mistrust and fear this level and type of corporate involvement, while corporations are frustrated with school bureaucracy and changing priorities and policies (Blair, 2000). However, communal forums that link schools and larger community development initiatives can help relieve some of these tensions.

If principals are to become builders of civic capacity, administrative preparation programs will need to be modified. Training programs will need to place greater emphasis on social justice, child development, collaboration, implications of poverty, and mobilizing community resources than they currently do. Recent standards for principal preparation fall short on these issues. While they emphasize knowledge, principals must have related to teaching and learning, they fail to develop the importance of the broader context in which principals must operate to establish true linkages within the community. As evidenced by reports from the principals in this sample, training programs need to help reshape how principals view their roles. Clearly, the urban principals in this study allocate their time primarily to functions internal to the school, and would allocate even more time in the same way if it were available. Leadership preparation programs must challenge the beliefs of principals about effective schooling.

Finally, thinking about the resilience of the problems civic capacity is attempting to address has implications for principals and the perception of education in general. These are tough problems, and they will require monumental investments from all. Despite such a commitment, solutions may not be found; progress will be slow. This may have negative implications for principal efficacy, which could result in reduced professional commitment and increased burnout. At a time when many argue a principal shortage exists, this could actually further reduce the candidate pool. Moreover, if the social ills schools and community partners are expected to cure persist, schools are likely to receive more than their fair share of the blame, which could further lower public satisfaction with public education. This decreased satisfaction could lead to more students going to private schools, being home schooled, or enrolling in other schools of choice, thereby leaving schools with fewer resources and higher percentage of students needing additional services. To prevent decreased principal efficacy and burnout, principals need to be taught about and spend time on self-care. Frank conversations about the challenges of education need to occur, not to place blame but to generate a greater understanding of the complexity of education. The overriding goal is developing a culture in which all community stakeholders share the responsibility of community well being.

REFERENCES

Beck, L. G., & William F (1999). Administration and Community: Considering Challenges, Exploring Possibilities. In: J. Murphy & K. S. Louis (Eds), *Handbook of Research on Educational Administration* (pp. 337–358). San Francisco: Jossey-Bass.

Blair, J. (2000). Corporations, educators work on strategic giving. *Education Week*, May 17, p. 12.

Bronfenbrenner, U. (1979). *The ecology of human development*. Cambridge, MA: Harvard University Press.

Brown, L. D. (1998). Creating social capital: non-governmental development organizations and intersectoral problem solving. In: W. W. Powell & E. S. Clemens (Eds), *Private Action and the Public Good* (pp. 228–241). New Haven: Yale University Press.

The Carnegie Foundation for the Advancement of Teaching. (1992). School Choice: A special report. Princeton, NJ: Author.

Crow, G. (1998). Implications for leadership in collaborative schools. In: D. G. Pounder (Ed.), *Restructuring Schools for Collaboration: Promises and Pitfalls.* (pp. 135–155). Albany, NY: SUNY Press.

Crowson, R., & Boyd, W. (1993). Coordinated services for children: Designing arks for storms and seas unknown. *American Journal of Education, 101*(2), 140–179.

Cumo, A, & Glickman, D. (1998). *What Works! In the empowerment zones and the Enterprise communities,* Vol II. Washington, DC: U.S. Department of Housing and Urban Development.

Dryfoos, J. (1994). *Full-Service Schools.* San Francisco: Jossey-Bass.

Empowerment, A New Covenant with America's Communities: President Clinton's National Urban Policy Report. (1995). Washington, DC: U.S. Dept of Housing and Urban Development, Office of Policy Development and Research.

Gittell (1994). School reform in New York and Chicago: Revisiting the ecology of local games. *Urban Affairs Quarterly, 30*(3), 136–151.

Honig, M. I. (1998). Opportunity to lead for school-community connections. Paper presented at the Annual Meeting of the American Educational Research Association. San Diego, CA.

Kirst, M. W., McLaughlin, M., & Bassell, D. (1989). *Rethinking children's policy: Implications for educational administration.* Center for Educational Research at Stanford, College of Education, Stanford University, CA.

Lawson, H. A. (forthcoming). Two new mental models for schools and their implications for principals' roles, responsibilities, and preparation. *Bulletin*, National Association of Secondary School Principals.

McLaughin, M. W. (1994). *Urban sanctuaries: Neighborhood organizations in the lives and futures of inner-city youth.* San Francisco: Jossey-Bass.

Pounder, D. G. (Ed.) (1998). *Restructuring Schools for Collaboration.* Albany: State University of New York.

Pounder, D. G., Ogawa, R. T., & Adams, E. A. (1995). Leadership as an organization-wide phenomena: Its impact on school performance. *Educational Administration Quarterly, 31*(4), 564–588.

Orr, M. (1996). Urban politics and school reform. The case of Baltimore. *Urban Affairs Review, 31*(3), 314–345.

Rallis, S., & Goldring, E. B. (2000). *Principals of dynamic schools.* Thousand Oaks, CA: Corwin Press.

Sergiovanni, T. (2000). *The Lifeworld of leadership.* San Francisco: Jossey-Bass.

Stone, C. N. (1998a). Civic capacity and urban school reform. In: C. N. Stone (Ed.), *Changing Urban Education* (pp. 250–274). Lawrence, KS: University Press of Kansas.

Stone, C. N. (1998b). Linking civic capacity and human capital formation. In: M. J. Gittell (Ed.), *Strategies for School Equity* (pp. 163–176). New Haven: Yale University Press.

Task Force on Desegregation of the St. Louis Public School System (1995). A report from the Civic Progress Task Force on desegregation of the St Louis Public School system St. Louis: Author.

Timpane, M., & Reich, B. (1997). Revitalizing the ecosystem for youth. *Phi Delta Kappan, 78*(6), 464–470.

Walsh, J. (1997). *Stories of renewal: Community Building and the future of urban America*. New York: The Rockefeller Foundation.

SCHOOLS IN THE BOWLING LEAGUE OF THE NEW AMERICAN ECONOMY: THEORIZING ON SOCIAL/ECONOMIC INTEGRATION IN SCHOOL-TO-WORK OPPORTUNITY SYSTEMS

Hanne B. Mawhinney

BACKGROUND: FROM SCHOOLS IN COMMUNITIES TO SCHOOLS GENERATING COMMUNITY

In 1992, Robert Crowson introduced the analyses presented in his book: *School-Community Relations Under Reform*, by challenging the prevailing view that schools are peculiarly 'intractable' institutions.[1] Crowson argued that "although intractability – that is, a unique capacity to remain unchanged by reform – may be a decided feature of the public schools, it is the central thesis of this book that something important and fundamental is currently taking place in discussions of school community relations" (pp. 2–3). Crowson also noted that the book "was written while efforts to restructure the public schools and to redefine school-community relations were well (indeed furiously) underway, but by no means played out. Some of the innovations such as home-schooling and business partnerships with the schools already have a bit of history; but many others (e.g. parental governance at the school site, and integrated services for children) are in the early stages of development" (pp. 3–4). Among the "early indicators" of the nature

Community Development and School Reform, Volume 5, pages 211–244.
Copyright © 2001 by Elsevier Science Ltd.
All rights of reproduction in any form reserved.
ISBN: 0-7623-0779-X

community-relations effects, issues, strategies, and alternatives, that Crowson reported[1] anticipated a number of new directions for school outreach:

> Schools and school districts under reform are increasingly turning the "old" direction of community and parental involvement in education around by engaging the schools and their professionals in programs of outreach toward the community . . . the relationships tend now to go well beyond PR toward some 'community-generating' endeavors between school and environment, toward a greater recognition of shared responsibilities for the welfare of children, and toward a 'new vocabulary' of professional cooperation in the delivery of services to children (p. 249).

Crowson was commenting on an emerging policy agenda for outreach with "a service-coordination flavor – a mission rooted in understandings of the importance of shared investments, integrated child development, and the value of mutually reinforcing interventions in the lives of children" (p. 249). The agenda loosely framed as the coordinated services movement did, indeed, generate new sets of relationships between schools and communities, however, subsequent research has shown they have been fraught with institutional tensions. Indeed the various reform efforts that were underway in the early 1990s have generated a new body of research and conceptualizations of the nature of the engagement of schools in their communities.

Mertz and Furman (1997) identify three general foci for the efforts that have been taken to create school-community connections: efforts to increase community control over schools through various forms of site-based management; attempts to coordinate services offered by schools and other agencies; and activities designed to engage parents in their children's education. Others have identified choice opportunities as a strategy for increasing connections between schools and communities (Murphy, 1999).

Research on each of these directions has shown that all are fraught with tensions, contradictions and complexities. Most research on site-based management suggests that rarely has this mechanism strengthened the broad-based links between schools and their communities (Malen, Ogawa & Kranz, 1988; Beck & Murphy, 1996). Research on efforts to coordinate services offers a paradoxical picture of the increasing engagement of schools with human services agencies, combined with an expansion of new professional and often bureaucratic roles (Crowson & Boyd, 1996; Knapp, 1995; Mawhinney, 1996; Smrekar & Mawhinney, 1999; Smrekar, 1996). Similar evidence of professional capture and control can be found in research on the engagement of schools with parents (Henry, 1996). Finally, those who have critically examined the potential of choice opportunities to reconnect schools to communities report that the interplay between choice and the cultivation of community is complex and often tenuous (Smrekar, 1996).

While this research certainly supports the intractability of schools hypothesis, other pressures have created a context which highlights the role of schools in community development. Crowson (1992) anticipated this new direction by noting that schools must become "community generating" institutions. However, the coordination of services vehicle for community generation that he anticipated proved to be less pressing than what some have called a "quiet revolution" in school to work transitional programs.[2] A key outcome of the revolution has been the School-to-Work Opportunities Act (STWOA) (1994). This legislation provides both a framework and an opportunity for states and local communities to design and implement different strategies to support students to master academic and technical skills, and prepare for further education and careers. The legislation provided funding for seven years during which time states received "seed" money to put their School-to-Work (STW) systems in place.[3]

Less than a decade after anticipating new demands on schools to become community generating, Crowson with colleagues Wong and Aypay (2000) argued that the school-to work revolution "seems to fit well into a recognition that the economic well-being of the entire community is fundamentally connected to the viability of public schooling" (p. 243). With this recognition they turned our attention to the need for policy theorizing on fundamental issues such as "*mobility* (denied or constrained) and *who's in control here* that could easily return to doom the revolution" (p. 248).

Such theorizing, they claim, has begun with the work of Mary Driscoll and Charles Kerchner (1999) who examined the potential integration of social and economic forces as schools engage with their communities. Drawing on the work of James Coleman (1990) and Pierre Bourdieu (see Bourdieu, 1977; Bourdieu & Passeron, 1977), Driscoll and Kerchner (1999) extended the concept of social capital formation to describe schools as contributing to the "revitalization of American cities" (p. 239).[4]

For Crowson, Wong and Aypay (2000), the value of Driscoll and Kerchner's (1999) theorizing lies in the construct that they employ to "view communities as zones of production-wherein the social forces associated with developing cultural capital (with schools as key institutional players) and wide-ranging economic forces (from job creation and economic investment to civic leadership (can come together in innovatively associated sets of relationships" (Crowson et al., 2000, p. 249).[5] At the same time, Crowson and his associates point out that "frankly, a deeper level of theorizing about the educational policy and administration implications of social and economic integration is largely yet to come . . . Attention must be given to what it means to engage in social and economic integration at the school/community level and how this goal is to be achieved" (p. 249).[6] They further argue that such analysis provides "an opportunity available to educators as never before to blend a rethinking of and a reemphasis on the school-to-work transition into a movement of significance in community-level revitalization in America" (p. 253).

Focus of Inquiry

With these comments as an impetus, in this chapter I examine the challenges that must be overcome if current initiatives supported under the School-to-Work Opportunities Act (1994) are to foster community revitalization. The case of current efforts to create school-to-work systems turns attention to the role that schools play in integration of the social and economic dimensions of civil society. In this chapter I argue that theorizing on the nature of this role must take into account ideas about the value and nature of such integration that have gained current public legitimacy.

I am among those analysts who believe that any attempt to theorize on the implications for education policy and administration of social and economic integration at the school/community level must take into account the broad-based ideas that frame discourse on the nature of such integration. Recognition of the role of ideas in determining how policy and administration is constructed in a particular domain has broad support among political scientists (see Braun & Busch, 1999; Reich, 1988; Sabatier & Jenkins-Smith, 1993; Sabatier, 1999; Stone, 1988). Theoretical approaches to understanding how ideas influence policy change have been developed and used to analyze policy changes in a range of arenas and have resulted in a growing base for rich theoretical debate.[7] In this chapter I contribute to this base by examining the ideas underlying efforts to forge social and economic integration at the school/community level through the development of school-to-work systems across the United States.

I argue that theorizing on the implications for education policy and administration of such integration can be framed by two ideas that have gained currency. The widely supported elements incorporated into the School-to-Work Opportunities Act (1994) (STWOA) reflect ideas about the integration of the social and economic that have both current currency and a long history. I show that Dewey (1916b, 1950) set out a grounding framework for what such integration means. I also argue that the distinguishing feature of the STWOA, that is its support for local community development of opportunity systems, demands that we examine the implications for such development suggested by another public idea of currency. That idea is embodied in Putnam's (1993a, b, 1995, 1996) claim that civil society in America has been weakened by the erosion of networks of relationships built on trust and reciprocity.

I argue that new approaches to community economic and social development that are being developed in response to the conditions of weakened civil society described by Putnam can guide the development of local opportunity systems for school-to-work transitions that contribute to community revitalization. Examining such developments offers one approach that can be used to respond to questions posed by Crowson and his colleagues about what it means to engage in social and economic integration at the school/community level and how this goal is to be achieved.

I begin my exploration by considering the implications for new roles for schools as part of developing communities that flow from the public ideas embodied in the systems being developed for school-to-work transition, and the current emphasis on developing civil society through networks of relations that build social capital. The first section of the chapter introduces each of these public ideas. A second section of this chapter examines new approaches to community development that build on Putnam's conceptions of social capital. In the final section of the chapter, I explore one of these implications: the need to adopt approaches to forming school-to-work partnership systems that develop the kind of networks of trust and reciprocity that build productive social capital in communities.

SECTION I: PUBLIC IDEAS ABOUT THE INTEGRATION OF SOCIAL/ECONOMIC CAPITAL IN COMMUNITIES

Two ideas have captured the attention of the public in recent years. One idea in currency emphasizes the need for America to recreate a sense of civil society to overcome the loss of social capital evident in the erosion of networks and norms of civil engagement. A second idea calls for new approaches to fostering the transition of all youth from school to work. Both ideas have achieved the status of 'public ideas,' that is they have created "a reason for someone to take action by setting out public value or necessity of the act and by giving the action a social meaning that is accessible to both the person who takes the action and others who are its audience or object" (Moore, 1988, p. 75). Both can be seen as public ideas because they have mobilized diverse interests in a common dialogue that has more or less coalesced into action agendas for change in how we view the relationship of schools to their communities. At issue is the role of schools in mediating the integration of the social and economic capital of the communities they serve. This chapter explores the impetus that the dialogues surrounding these two public ideas have provided for rethinking the role of schools in developing communities.

The Search for Civil Society

In its July 22, 1996 issue, *Time* ran a short article discussing the merits of the idea of social capital as described by Robert Putnam, a Harvard political scientist. (Stengel, 1996). Referring to Putnam's analysis reported in an essay entitled "Bowling Alone", *Time* journalist, Richard Stengel, commented "Putnam's essay seemed to reinforce a widespread feeling that civil life in America wasn't what it used to be ... The nation's diminished social capital was widely lamented far and wide, from Bill Clinton's bully pulpit to angry sermons by Bill Bennett ..." (1996, p. 35). Putnam (1995) described the loss of social capital in American life through slow but steady decline in civic participation. He concluded that without participation in public life, feelings of trust and bonds of reciprocal help are undermined, ultimately weakening the ability of communities to solve problems and sustain economic prosperity.

The fact that the idea of social capital had reached the readers of *Time* underscored its status as a public idea, providing a rationale and supporting value for action.[8] Indeed, its status as a public idea of this sort was already evident when *Time* published Stengel's commentary. A few months earlier, *Atlantic Monthly* had also featured social capital as a new policy idea emerging at a time when traditional theories of economic and social change have largely proven inadequate. In the *Atlantic Monthly* article, Nicholas Lemann concluded that "'Bowling Alone' struck a nerve in part because it provided a coherent theory to explain the dominant emotion in American politics: a feeling that the quality of our society at the everyday level has deteriorated severely" (1996, p. 25).

Putnam is, of course, not the only social scientist to examine the nature of social capital. The concept is an element of Pierre Bourdieu's (1986) general theoretical approach to analysis of social life. It is a particular formulation of 'capital', which provides the fuel for social mobility. Bourdieu suggests that capital exists in various forms of resources: human, financial, cultural and social, that are used for strategic action. In material or symbolic forms capital provides resources or access to resources that can enhance mobility. In Bourdieu's analysis social, cultural and economic capital overlap and inter-connect, moreover, the possession of one form of capital can reinforce the power of another, or the capacity to acquire another. Social capital in Bourdieu's analysis is "the aggregate of the actual or potential resources which are linked to the possession of a durable network of more or less institutionalized relationships of mutual acquaintance and recognition – or in other words, to membership in a group-which provides each of its members with the backing of collectively-owned capital, a 'credential' which entitles them to credit, in the various senses of the word" (1986, p. 248–249).

The relational nature of social capital is also underscored in the conception offered by James Coleman (1990). In Coleman's analysis, social capital is produced through a wide variety of social relations, it "inheres in the structure of relations between persons and among persons"(p. 302). Its presence helps to facilitate both beneficial and harmful interactions. For Coleman, social capital that benefits a narrowly defined social group may not benefit a larger social group or society in general. Coleman's work has set important new directions for researchers concerned with the nature of the engagement of schools with communities, as will be described in sections of this chapter to follow. It is, however, Putnam's (1993a, b, 1995) conception of social capital that will be developed in this chapter. In this conception social capital takes on a broader role but has a narrower meaning than it receives from Coleman (1990). Its relevance to this chapter is its focus on the development of civil society, a focus that draws attention to the particularly challenging issues of local community development, that are also central to the action agenda for promoting school-to-work transitions (the second public idea considered in this chapter). Social capital from this perspective requires that the social structures developed in support of school-to-work transitional activities be oriented toward promoting positive civic action and economic cooperation in local communities.

A Public Agenda for Local Community Development of School-to-Work Systems

In recent years a school-to-work movement has emerged reflecting consensus cutting across political parties and regions of the country that attention must be paid to the transition of youth to careers. An initial impetus for this concern came from reports outlining the experiences of those youth that graduate from high school but do not pursue a baccalaureate degree. Reports referring to this group as the "missing middle" (Thurow, 1985), were followed by the influential report issued by the William T. Grant Foundation (1988), referring to the "forgotten half." As an initial focus for action this group of youth were logical targets for action of a school-to-work movement that had its origins in public debates surrounding the 1993 publication of *A Nation at Risk* (National Commission Educational Excellence). That report had focused intense national attention on the shortcomings of the American educational system and set the stage for federal and state reform efforts. Its central premise was that the nation's preeminence in science, technology, industry, commerce, and military preparedness was threatened by its mediocre primary and secondary schools.

Two other national reports took these concerns as points of departure and further focused public attention on the issues of transition to work:

America's Choice: High Skills or Low Wages (Commission on the Skills of the American Workforce, 1990), and *What Work Requires: A SCANS Report for America 2000* (Secretary's Commission on Achieving Necessary Skills, 1991). These reports reflected a broad-based consensus about the nature of the transitional links between schools and work that best respond to the needs of American communities.[9] Elements of consensus around these opportunity systems were described by Robert Reich, who as the 22nd Secretary of Labor in the Clinton administration oversaw with Robert Riley, the Secretary of Education, the development of the STWOA. In an essay describing the building of a framework for a STWO system, Reich (1995) described the expectations that states and local communities would be able to create statewide school-to-work opportunity systems for students by:

- Building partnerships among schools, employers, labor, community organizations, and parents to develop and sustain school-to-work opportunity systems as part of a lifelong learning system for the nation;
- Enlisting employers in providing work-based learning opportunities to young people as part of their high school experience; and
- Reforming secondary schools and programs for out-of-school youth (pp. 128–129).

Evident in these elements is a "public idea" that *local community-level partnerships* (emphasis added) must be built, as Reich noted, "to develop and sustain school-to-work opportunity systems as part of a lifelong learning system for the nation" (Reich, 1995).[10]

Given the broad agreement on the fundamental value of the idea of local control in the design of the Act, it is not surprising that current reports on the implementation of the STWOA offered as evidence of the success of the seed funding provided to states reflects the wide diversity of the initiatives.[11] Indeed two recent reports present such diversity as evidence of success. The Secretaries of Education and Labor, the two federal government departments responsible for the School-to-Work Opportunities Act (STWOA) of 1994 presented one report to Congress in 1998. The Secretaries reported that the goals of the Act are ambitious and include the following:

- Helping students achieve high-level academic and occupational skills;
- Widening opportunities for all students to participate in post secondary education and advanced training, and move into high wage, high-skill careers;
- Providing enriched learning experiences for low-achieving youth, school dropouts, and youth with disabilities; and assisting them in obtaining good jobs and pursing post secondary education;

- Establishing the framework in which all States can create School-to-Work systems that are part of comprehensive education reform and career preparation;
- Increasing opportunities for minorities, women, and people with disabilities, by enabling them to prepare for careers from which they traditionally have been excluded; and
- Utilizing workplaces as active learning environments.[12]

The Secretaries then reported "meaningful progress in the implementation of the STWOA that addresses these goals and reflects local needs and decision making. The data discussed in this report indicates that STW systems are building a great variety of State and locally-designed educational and career activities and services and that these are available to increasing numbers of students at all levels of schooling."[13]

Two years later, comments by the Director of the STW Opportunities Office on the implementation of these goals were celebratory. On October 11, 2000 at the School-to-Work 2000 Conference on Capital Hill in Washington, DC, Stephanie Powers, the Director welcomed participants by stating "we have much to celebrate as we look back over the past $6^1/_2$ years since the School-to-Work (STW) Opportunities Act was passed in 1994!"[14] She went on to list as causes of celebration evidence that:

- States have taken up the challenge of sustaining initiatives that were provided start up funding by the STW Opportunities Act by increasing to a total of $90 million allocations to these initiatives.
- That more than half of STW initiatives are now funded from other sources.
- That STW is supporting students attainment of high academic standards because the percentage of students in local STW partnerships taking at least three years of math and science is increasing and students participating in work-based learning in some communities are earning higher grade point averages.
- STW is working is working for business: the numbers of employers offering work-based learning opportunities has risen from 60,000 in 1995–'96 to 150,000 in 1998–'99.
- STW strategies have influenced initiatives in the Federal Departments of Education and Labor, including the Labor Department's youth opportunities movement and new youth councils; and in the Department of Education, the new Small Learning Communities initiative, the reauthorization of the Perkins Act, and now STW strategies underpin the emerging high school reform movement.[15]

A concluding piece of evidence of success of the STW initiative that was offered by Powers articulates some elements of this debate. Powers concluded her list of evidence of success by speaking directly to those she called the "contrarians of STW, saying "we now know for sure that STW broadens student's options for careers and college, not limits them, as they so effectively claimed before STW was even implemented. Well, the proof is in, and . . . they were just plain wrong!"[16] In referring to 'contrarians', Powers described the critics of the STWOA, some of whom were skeptical of the potential of STW initiatives to address the social dysfunctions created by the old-style vocational tracking system that evolved from the vocational education movement active from 1900 to 1917.[17]

Skepticism and Support for STW Opportunity Systems

One of the issues of concern for current critics was, and continues to be, whether the STWOA can take into account the need to ensure that any efforts to improve transitions from school-to-work are framed to support values of socioeconomic justice, and that they do not simply serve the new global economic dynamics. As expressed by Kincheloe (1999), these critics wish to ensure that the innovations proposed by STWOA reflect an orientation towards helping "future workers discern the difference between a genuine worker empowerment and corporate lip service to empowerment" (p. 410). Kincheloe argues that "such corporate pseudo-empowerment claims that workers are wanted who are critical and can think for themselves; upon further examination, however, we find that workers in many of these 'empowered' workplaces are limited in the actions in which they can engage, the types of options they can choose, and the prerogatives they can exercise. The managerial message is confusing: 'Be empowered, think for yourself, but don't let it get out of hand'." (p. 410).[18]

Supporters of the STW systems being developed argue that these reforms are intended to address the dysfunctions of vocational tracking that so concern critics like Kincheloe (1999). Indeed they point out that it is noteworthy that not only was there broad bipartisan support for the Act in both the U.S. House of Representatives and in the Senate,[19] but that as it has been implemented, it has continued to garner support from education, labor, business and community groups. There is broad-based support from for an Act that addresses some of longstanding issues surrounding vocational versus liberal studies that have divided scholars of education since the debates between Dewey and the social efficiency philosophers of the early 20th century (Wirth, 1991).

The debates are perhaps important to consider when assessing the potential of STW systems being developed today to address the concerns those earlier debates reflected. By Wirth's account the social efficiency philosophy expressed by Charles Prosser who eventually became the executive director of the Federal

Board for Vocational Education saw a technocratic model of vocational education as essential to address human capital demands facing the new manufacturing economy of the country. This model was marked by "a conservative social philosophy, a methodology of specific training operations based on principles of S-R psychology, and a curriculum designed according to a job analysis of the needs of industry, and by a preference for a separately-administered set of vocational schools" (Wirth, 1991, p. 56). The underlying philosophy was expressed by Prosser who argued that vocational education "only functions in proportion as it will enable an individual actually to do a job . . . Vocational education must establish habits: habits of correct thinking and of correct doing. Hence, its fundamental theory must be that of habit psychology" (Prosser & Quigley, 1950, pp. 215–220). By 1917, Prosser's conception of vocational education had been adopted by policymakers, creating a heritage that current critics like Grubb (1995) argue resulted in the "a series of dismal consequences" that spun out from the separation of academic and vocational purposes of education:

> Even though academic courses remained the choice of most students, it was always clear that the majority would not go to college or enter the professions, so the content became slowly degraded for most students – particularly in the courses of the general track. The purposes of the curriculum – its original link to college preparation and then to professional occupations – became increasingly murky for the majority of students; the link to future work – the reasons why the competencies learned in school might be important in later life – became increasingly vague. The links to institutions outside the school, never strong to begin with, continued to erode as the rationale for working with employers and community organizations weakened. And the need to differentiate students 'scientifically.' With some destined for the professions while the rest prepared for semiskilled and unskilled jobs, in turn led to the current apparatus of testing and tracking students (p. 2).

In part the impetus for the efforts to design a new STW system grew out of broad-based awareness that these dysfunctions were reflected in growing evidence of the disconnect between the world of school and life after and outside school, the domination of "academic" instruction, and the emphasis on the college-bound along with the neglect of the "forgotten half" not bound for college (W.T. Grant Commission, 1988). It is noteworthy that Dewey's (1915, 1916a, b) critique of the social efficiency philosophy that founded vocational education in the U.S. provides a grounding theoretical framework upon which many reformers have built their proposals for the essential elements of a new STW system.[20] Most important, for the purposes of the focus of this chapter, Dewey offers a framework from which to examine the potential for schools to mediate social/economic integration through their efforts to support the range of initiatives that have come to be associated with the School-to-Work Opportunities Act (STWOA).

Dewey's Framework for Education for Democracy

Dewey's criticism of the social efficiency philosophers' proposals for a new vocational education underscores the critical stance that must embed the roles that schools take in mediating social and economic integration. Dewey wrote in 1915:

> The kind of vocational education which I am interested in is not one which will 'adapt' workers to the existing industrial regime; I am not sufficiently in love with the regime for that. It seems to me that the business of all who would not be educational time-servers is to strive for a kind of vocational education which will first alter the existing industrial system, and ultimately transform it (1915, p. 42).

Dewey's conception of the nature of the individual and the problems of democratic traditions in a technological society framed his concerns. Wirth (1991) argues that in Dewey's view "if you wanted persons with qualities capable of sustaining democratic values, they had to be nourished in communities marked by such values" (Wirth, 1991, p. 61). The problem for Dewey was whether democratic values of meaningful participation could be sustained under conditions of a new economic order marked by urban-corporate pressures. Even when he wrote, the contradictions between the rhetoric of democratic values and the lived experiences of workers were manifest. For Dewey the task was "to develop strategies for bringing qualities of the democratic ethos into institutions being transformed by science, technology, and corporatism" (Wirth, 1991, p. 61). Schools played a critical mediating role between the social and the economic.

Underlying his depiction of this role was Dewey's critique that the use of science and technology for the single-minded pursuit of profit reflected in a laissez-faire economy was flawed and dangerous. Children and youth had to be educated to understand how to examine the consequences of technology. Schools played a key role providing youth with an understanding of the social (and democratic) consequences that must be considered in using science for economic development. He viewed education as involving more than training. As one of the vehicles of education, the test of the value of schools rests on their capacity to develop in children and youth the ability to examine the consequences of technology for social life. The criteria by which judgments can be made of the value of the mediating role between the social and economic that school can play are articulated by Dewey in the following often cited statement in the *Reconstruction of Philosophy*:

> All social institutions have a meaning, a purpose. That purpose is to set free and to develop the capabilities of human individuals without respect to race, sex, class or economic status . . . [The] test of their value is the extent to which they educate every individual into the full stature of [his] possibility. Democracy has many meanings, but if it has a moral meaning, it is found in resolving that the supreme test of all political institutions and

industrial arrangements shall be the contribution they make to the all-around growth of every member of a society (Dewey, 1950, p. 147).

According to Dewey, all-around growth was best fostered by "education through occupations [not for occupations because it] combines within itself more of the factors conducive to learning than any other method" (1916b, p. 309). His rationale rested on two questions that he saw as fundamental: "Whether intelligence is best exercised apart from or within activity which puts nature to human use, and whether individual culture is best secured under egoistic or social conditions" (Dewey, 1916b, p. 320). Dewey's responses consistently framed the conditions of effective learning as occurring in a social context. For Dewey, then, education through the occupations implied not only an opportunity to provide a social context for learning; it addressed deeper issues surrounding the dilemmas of integrating the social and economic spheres. Dewey argued that educating through the occupations was a good method for organizing instruction because "An occupation is the only thing which balances the distinctive capacity of an individual with his social service" (Dewey, 1916b, p. 309). By this Dewey pointed out that education through the occupations requires that schools find ways of balancing the individual interests of youth in developing capacities and finding possible roles as adults, with the interests of communities in having well trained and productive workers.

Dewey's Influence on STW Transition Designs

Dewey's framework for thinking about the integration of the social and economic by education through the professions did not prevail. However, it has become a guide for some those who have lead a movement to integrate academic and vocational strands of schooling. Researchers like Grubb (1995) have used Dewey's framework as a rationale for approaches to integration that have been incorporated into the new STWOA as fundable options. Contributors to Grubb's two-volume examination of education through the occupations described an array of approaches. These include the integration of academic content into vocational courses, involving academic teachers in vocational programs to enhance academic content in vocational programs, making academic courses more vocationally relevant, alignment of vocational and academic courses, the development of senior projects which focus on transitions, developing academies which may take the form of schools-within-schools, the development of occupational high schools and magnet schools where courses are aligned and take an occupational focus; the development of occupational clusters, career paths and majors (Grubb, 1995, pp. 59–96). In fact, many of these approaches are embedded in the three key components of the framework established by the STWOA (1994). These components include:

(1) School-Based Learning. Rigorous classroom instruction that is linked to workplace experiences and that provides students with the information and skills needed to identify and prepare for promising careers.
(2) Work-Based Learning. Work experience, structured training, and other workplace activities appropriate to student's career interests and linked to their school curricula.
(3) Connecting Activities. Efforts undertaken by partnership members to help employers and schools forge and maintain links between the school-based and work-based components (Hulsey, Van Noy & Silverberg, 1999, p. 1).

The fact that the STWOA supports these components of current transition systems does not ensure the realization of the kind of democratic outcomes of ensuring the "all-around growth of every member of society" envisioned by Dewey. These components are embedded in a design based on the values supporting local control while acknowledging the diversity of outcomes resulting from such control. Indeed, the public and nonpartisan political agreement that produced the STWOA was based on an American belief that local communities (with states as guardians) are the legitimate site and initiators of revitalization of social and economic systems of opportunity. When translated into a public idea about effective school-to-work opportunity systems, a recent national evaluation of STWOA finds that "one of the most distinguishing characteristics of the National School-to-Work initiative is the discretion left to states and localities to define their goals and develop strategies for achieving their goals" (Erlichson & Van Horn, 1999, p. 4). The report notes while the community focus is appropriate, it is not without problems. Among these the report notes "most important is that there is not a standard definition of what School-to-Work is or what the goals of the initiative should be. There are multiple names for the initiative both among the states and sometimes even within a single state" (p. 4).

Conceptualizing Schools Engaged in Community Revitalization

Putnam's (1993a) arguments suggest that local diversity is not a problem if the transitional systems are being built through connecting activities that develop the kinds of local networks linking school, business, and community that create social capital. We know however, from studies of the early implementation of

STWOA funded initiatives that such connecting activities are far from well developed.[21] It is clear that challenges that must be addressed in designing connecting activities reflect differences in the governance structures adopted by different states. These governance structures involve "decisions such as where to locate the state STW office, how to define the interaction between the state office and the sub-state partnership, how to garner support and participation from other state-level agency partners, and how to garner support and participation from non-governmental partnerships such as business and industry" (Erlichson & Van Horn, 1999, p. 27). The arguments of Dewey and Putnam suggest that such decisions should be made based on principles of developing civil society in communities. Moreover, returning to the challenges that Crowson, Wong and Ahayet (1999) identify, such decisions required that schools rethink the means by which development can foster the integration of the social and economic.

These decisions must be based on broader conceptions of schools engaged in community revitalization than are currently held. It means that local partnerships formed as a "connecting activity" under the STWOA parameters for funding, should be framed by models such as Boyd, Crowson and Gresson (1997) describe as 'enterprise schooling'. In such models a local school would "be a fully active player in a developmentally oriented network of public/private community institutions from banks, to churches, to employers, to 'activists' " (p. 92).

Many of those who have begun to conceptualize what such enterprise schooling requires have used Coleman's (1990) conceptions of social capital as a point of departure.[22] Coleman defines social capital by its function, not as something possessed as a single entity. Rather, it is "a variety of entities having two characteristics in common: They all consist of some aspect of a social structure, and they facilitate certain actions of individuals who are within the structure. Like other forms of capital, social capital is productive, making possible the achievement of certain ends that would not be attainable in its absence" (p. 302). Social capital therefore, resides in relationships, both those within an organization, and those between individuals. Coleman argues that the formation and maintenance of social capital also depends upon the reinforcement of a normative frame that emphasizes the public good aspect of group relations and indicates "the importance of every member of the group as a whole" (Coleman, 1990, p. 321). Finally, Coleman explains that:

> Like human capital and physical capital, social capital depreciates if it is not renewed. Social relationships die out if not maintained; expectations and obligations wither over time; and norms depend on regular communication (p. 321).

These explanations of the factors that impact the creation, maintenance and destruction of social capital form a theoretical scaffold from which a new generation of scholarship is examining schools as "coherent sets of relationships . . . ordered by trust, knowledge and authority" (Driscoll & Kerchner, 1999, p. 390). Driscoll and Kerchner argue that this new exploration requires that we examine the interpersonal relationships among all constituencies as well as the nexus between school and community" (p. 390).

In an examination of the theoretical challenges these arguments pose, I have argued elsewhere that an ecological perspective on the nature of these micro-relations should frame such exploration. Using a microecological framework of analysis, I show in a case study of a high school engaged in 'enterprise schooling' that "the relationships created of trust, legitimately perceived authority, and transfer of knowledge that are fostered between students, school personnel and members of their school communities enhance the flows of social capital not only for the benefit of individual students, but, as well, for the public good of the whole community" (Mawhinney, in press).

While Coleman's (1990) conceptions have given impetus for important new theorizing on the nature of social capital flows, it is Robert Putnam's (1993a, b, 1995, 1996) concern with the nature of civil society in local communities that has captured the attention of community economic and social planners and developers. It is important to note differences between Coleman's claims for social capital and the use that Putnam makes of it. Coleman frames social capital as a component within a nested structure of social relationships and processes. He does not explicitly frame its operation within the context of civil society, as Putnam does. Analysts agree that Putnam's emphasis on the role of voluntary associations in its generation, and his interpretation of social capital in relation to civic engagement draws from Tocqueville's interpretation of democratic societies (Foley & Edwards, 1997) Putnam's analysis of the role of voluntary associations in the development of civil society, along with his interpretation of the role of social capital in explaining and predicting success in economic competitiveness have captured the attention of community/ economic development theorists, practitioners and advocates. I argue in the second section of this chapter, new theories of community development based on Putnam's analysis of the link between social capital formation and economic development are particularly important to take into account when considering the challenges posed in building local partnerships to develop STW opportunity systems.

SECTION II: THEORIZING COMMUNITY DEVELOPMENT AS SOCIAL/ECONOMIC INTEGRATION

STWOA Design Principles: Building Local Capacity

In the current policy environment the idea of social capital has energized new areas of focus for addressing public policy issues. *Atlantic Monthly* columnist Lemann observed that "if Putnam is right that as local problems go, so goes the nation, his work suggests the possibility of solving our problems through relatively low-cost association-strengthening local initiatives that don't require higher taxes" (1996, p. 5). Putnam's focus on local communities as the appropriate site for problem solution is not unique. Indeed one of the significant forces in changing the way government's 'do business' in recent years involved devolution to communities through "designed-in" support for local flexibility. This was one of the driving forces in the STWOA design. It is noteworthy that general concern with the state of workforce training and vocational education preceding the design of the Act focused on the dysfunctions of the system of uncoordinated and overlapping training programs that existed prior to its passage in 1994. Politicians involved in its passage note that the Act was part of the Clinton administration's agenda for "reinventing government" intended to support initiatives that would support both "local flexibility" yet provide "nation-wide consistency" (Reich, 1995, p. 125) Reich explains:

> We knew that we had to galvanize the full array of state and local job training and education resources. New federal resources would be limited ($100 million was ultimately appropriated for our first year of implementation in FY 1994). It also was important for us to craft a new kind of bill – one that reflected the spirit of 'reinventing government.' It could not be a typical top-down approach. To ensure that there would be both local flexibility and nationwide consistency, we developed a number of strategies that eventually were adopted by the Congress – and that now permits rapid construction of a new system (p. 125).

Reich listed several strategies that were intended to capitalize on local community capacity. States were given multiple avenues to use to build STW systems with federal support. Development grants, implementation grants and waivers to provisions in existing job training and education programs were made available to states. The goal was to "enable faster start-up and diffusion of STW systems and more flexible and creative strategies. Reich underscored the local capacity building goals by explaining that the STWOA provides venture capital for states and local communities to build a school-to-work system:

> Our goal is to promote ongoing community ownership of and responsibility for bettering young Americans' career and educational opportunities, but not to create another top-down

permanent federal program . . . By design we are leaving considerable room for experimentation and local diversity . . . [however] Business and labor leadership is critical to the success of all aspects of this initiative. Employers – in partnership with labor – will play a key role in the design and implementation of the system, including defining skill requirements for jobs, serving on state and local partnerships and governing bodies, offering quality work experiences for students, and providing post-program job opportunities for students and graduates (Reich, 1995, pp. 125–126).

Challenges in Forming STW System Partnerships

Reich underscored that partnership formation was intended to be the vehicle for achieving the STWOA's goals of promoting ongoing community ownership of, and responsibility for, bettering the educational and career opportunities for youth. STW opportunity systems being promoted in this design depend on the development of networks among community members representing business, labor, and education. Evidence in the report, *Making Good Jobs for Young People a National Priority* by the National Center on the Educational Quality of the Workforce (EQW), emphasized the importance of linkages between investments in education and training and the restructuring and reorganization of work as the American economy makes the transition to an economy founded on new production and employment systems (Saul, 1998). As a chairperson of the Advisory Committee to the EQW, Robert Saul pointed out that the *National Employer Survey* (NES) (1995) documented "the link between productivity and successful business adaptation, on one hand, and investments in education and training on the other" (Saul, 1998, p. 171). Despite the productivity payoffs for employers of increasing educational attainment of workers, the EQW-NES survey conducted in 1995, suggested that most do not see schools as effective partners in developing that workforce. The survey revealed that: few employers pay attention to measures of school performance when making hiring decisions, teachers recommendations are not generally considered, few consider grades as important, nor was the reputation of the school considered in hiring decisions by most employers. Saul argued that such evident disconnection will result in a "significant erosion of education's capacity to contribute to establishment productivity" (p. 171). He noted similar concerns from the education community, citing Albert Shanker's (1995) commentary in the *New York Times*:

It's obvious that the less attention employers pay to school performance, the less incentive kids have to achieve and the more poorly prepared they will be . . . Until Businesses can have confidence in student transcripts and recommendations, they will go on discarding them – and students will continue to conclude that what they do in school does not count (Shanker, 1995, April 23).

Saul further warned that the failures of high-level partnerships involving schools and business leaders compounded these problems. Writing in 1998, Saul noted that "many business leaders entered these partnerships with the best of intentions, but, by and large, the accomplishments of these partnerships have been marginal, and they have not tackled the systemic problems of urban public education" (p. 172). Referring to the diffuse goals and public relations orientation of past partnerships which had no effective mechanisms for monitoring and evaluation, the EQW Advisory Board made recommendations to improve the linkage between schools and employers based on 'a simple axiom: be practical and think locally" (p. 173). The Board called for concrete, low-cost initiatives-administered at the state and local levels-that foster better, more substantive exchanges between employers and schools" (p. 173). It called for employers to make use of grades and the reputation of a school as important criteria in hiring. Work-learning programs, such as apprenticeships and internships, were supported by the Board as a means of making explicit the connection between school and work. They have "the added advantage of creating an informal but effective communication channel through which employers learn about schools and their students and through which schools learn about the needs and practices of employers (p. 174).

The design of the STWOA incorporated the intentions evident in the report of the Advisory Board of the National Center on Educational Quality of the Workforce (1995). Local partnerships have been promoted by funding through the STWOA. The National School-to-Work Office reported in November 2000, that STW was being implemented in every state, and many states were moving towards sustainability. The Office reported that between 1995 and 2000 there was a 290% increase in the number of students in school in geographic areas served by local partnerships.[23] The Office also reported that employer involvement in STW was widespread and expanding, noting that the percentage of employers nationally involved in STW partnerships grew from 25% in 1996 to 37% in 1998.[24]

A New Socio-Economic Impetus for Revitalizing Civic Engagement

Some of these good news reports, however, contain warnings about the difficulty of achieving the goals of the STWOA which Reich (1995) called for that emphasized partnerships which promote ongoing community ownership of, and responsibility for, bettering young Americans' career and educational opportunities. There is general agreement that such goals require new ways of looking at what it means to engage in social and economic integration at the school/community level and how this is to be achieved. Certainly there is concern that the impetus for such integration will be undermined if initiatives cannot be sustained following the sunsetting of the STWOA in 2001. In her comments to the *School-to-Work 2000: Thinking About Tomorrow Conference*, the National Director of the STWOA Office reported that more than half of the STW partnerships around the country are receiving cash funds from other sources than the STWOA's funding. However, these percentages mask the extent of the effort that must be made if the problems that the STWOA targeted are to be addressed given the demographic changes underway in the U.S. In a 1998 report to the Organization for Economic Co-operation and Development (OECD) as part of the *Comparative Study of Transitions from Initial Work Life in 14 Member Countries*, the National Center for Postsecondary Improvement concluded that:

> The United States faces important demographic shifts. Beginning in the mid-1990s, the size of the cohort graduating each year from high school began to increase. Through the first decade of the twenty-first century, the number of young people making the transition from initial education to working life will increase each year. At the same time, the first of the 'baby boomers' will begin leaving the workforce, making the economy and the country increasingly dependent on young workers, whose efforts will have to fund the retirements of so many of their grandparents' generation (Zemsky et al., 1998, pp. 52–53).

At issue, is whether the partnerships initiated through the seed funding of the STWOA are robust and extensive enough to address the challenges that these demographic changes will bring to communities. Zemsky and his colleagues suggest "the looming crisis over Social Security in the United States is as much about economic productivity as it is about the actuarial soundness of the system itself" (1998, p. 53). From the vantage point of communities these changes suggest that STW opportunity systems should be developed through partnerships that reflect broad-based civic engagement in finding ways to build and sustain economic prosperity. In the concluding section of this chapter, I outline promising new directions in community development that target the development of the kind of social capital that is needed to address the challenges that communities will face in this new socio-economic context.

SECTION III: NEW DIRECTIONS IN COMMUNITY DEVELOPMENT THROUGH SOCIAL CAPITAL FORMATION

Roots of Community Development Practice

The literature on social capital argues that "networks of support and influence found in a community are functionally related to improving social conditions, both directly and through better implementation of social policy" (Edelman, 2000, p. 172). Implied here is a psychosocial conception of community that dominates current public and policy debates (Edelman, 2000). In this sense community is characterized as reflecting a sense of 'we-ness', of solidarity and of cohesiveness (Martinez-Brawley, 1995). Implicit in this view of community is a conception of the nature of modern society that has taken root as modern liberal democracies such as the United States have developed. Plant (1991) points out that "It is important to recognize that the idea of community has frequently been invoked over that past 200 years as an attempt to correct what has been seen as the individualism, the subjectivism, the atomism, the alienation, the instrumentalism, and the contract-based and market oriented character of modern society liberalism" (p. 325).

Underlying the search for community, and its development, is the idea that the layers of social organization that exist between individuals and large economic and political institutions are weak. A common logic following this idea is that weakened ties in families and communities have led to serious social and individual pathologies. Social policies and programs framed as developing community are typically designed to counter these conditions. The logic of community development is that the closeness and relatedness developed in a community enable people deliberate as citizens in solving problems.

Community development strategies presume that the community is a manageable site at which to provide a comprehensive response to social problems, while still being small enough to meet critical socio-pyschological dimensions that are the target of concern. Edelman (2000) explains that "contradictions that appear, for example, in large organizations (between accountability and efficiency), in mass politics (between legitimacy and administration), or in democratic theory (between liberty and equality) become manageable dynamics at the community level of organization because of the adjustments facilitated by face-to-face interactions" (p. 173).

Community development has long been associated with strategies for social change. The past twenty years community development approaches taken in the United States range across diverse ideological perspectives. Noteworthy is the mix of public and private domains of civil society that have poured billions of dollars, and multiple other resources, into a range of activities encompassing, voluntarism, public service programs, community partnerships, community based social services, faith-based alternatives, community-education programs, and intense community organizing (Edelman, 2000, p. 169).

Challenges of Community Development

In practice, the potential of community development oriented programs to affect change is limited. There is a long history of the failure of community-based efforts to reach their objectives. Analysts point out that "for each epoch of community action, from the Progressive Era through the New Deal, from the sixties' War on Poverty to the modern day, there are numerous analyses of failure" (Edelman, 2000, p. 169). Many observers of current efforts at community development take seriously the implications of assessments by those who studied the community initiatives of those earlier periods pointing to a long history of the failure of the strategy to meet expectations (Warren, 1975).

Researchers now ask whether these failures are due to a fundamental flaw in the concept of community development, or whether they are due to the challenges of implementing initiatives that take such an approach. Most evidence suggests that implementation challenges overwhelm possible positive outcomes. Schorr (1989, 1997), for example, argues that the problem is less a matter of lack of knowledge of principles of community development, than it is a lack of application of that knowledge.

Typically the language of community development holds a symbolic power to legitimize program activity. Language about mutual interaction, shared authority and responsibility, empowerment, inclusion and consensus that reflects a democratic mutualist organizational model support our psychosocial ideal of community. Most analysts agree, however, that this model of organization is more symbolic than reflective of the actual administration of community development efforts. In order to be implemented community-based programs must bring together, as a matter of both principle and strategy people from diverse social, economic and professional backgrounds. This raises difficulty because participants can have quite different substantive understandings of what community and related ideas mean and, as a result, may have quite different goals, strategies and agendas. Well-documented troublesome aspects of community development arise because of these differences among participants (Baum, 1999).

Often, an institutional community, convening of representative leaders from government, business, education, religious groups and other groups of civic importance, bring expertise, access and credibility that is often lacking at the community level. Bonds among these individuals are largely instrumental, reflecting the challenges to their collaboration created by the structures, interests, environmental controls, and conventions of institutions which they represent (Baum, 1999; Boyd & Crowson, 1996). Moreover, evidence from research on a wide range of community development-oriented programs underscores the power of traditional bureaucratic models to capture the strategic approaches used in implementation. Typically representatives of the institutional community choose strategies that fit the survival and growth needs of their agency or professional group. These strategies are typically framed abound formal rules, hierarchical organizational structures to control actions, specialization and professional expert control. Studies of coordinated service initiatives in education and other public service arenas have documented the disjunctures between the language of community used to legitimate programs, and the traditional bureaucratic practices of their administration. Strategic control by educators in coordinated services initiatives is well-documented (Mawhinney, 1996, 1998; Mawhinney & Smrekar, 1996; Smrekar & Mawhinney, 1999). Similar patterns of professional capture are reported in community development studies in health and human services domains (Edelman, 2000). Although partnerships may intend to hold to standards of inclusiveness, in practice many coalitions established to develop and govern community programs become less representative of a community and more representative of agencies and professional communities.

Social Capital and Community Empowerment

Despite these well-documented difficulties, the new concern with community revitalization has energized the further development of both long standing and emergent approaches to community development. The new focus on community empowerment moves beyond the emphasis on coordination, collaboration, and partnerships inherited from efforts to overcome the effects of disinvestment in public sector supports during the Bush and Regan regimes (Halpern, 1995). Implicit in the idea of social capital is a belief that local engagement can be used as component of community capacity-building. This is not a new concept, but in its current formulation it has given impetus for new support for Comprehensive Community Initiatives (CCI). These initiatives reflect a refocusing of attention on holistic responses rather than responses to individual issues such as unemployment, poor educational performance and the like (Wallis, Cocker & Schechter, 1998). CCIs instead, begin by focusing on the 'tangle of pathology' of which these problems are a part. An integral aspect of this new approach is the emphasis placed on working with and through communities, as the most effective way of addressing the needs of individuals.[25]

The approach has taken several thrusts, perhaps the most widely known application is captured by the idea of community 'empowerment'. Reflecting the translation of these ideas into policy, in 1994 the Clinton administration created Empowerment Zones (EZ) to aid distressed urban areas. This program provided federal funds while requiring cities to raise matching money.[26] Funding was to be accompanied by policy innovations, including tax incentives for development, and each EZ city was to designate an organization to manage policy, funding and programs. The central idea of EZ was however, that they would "catalyze systemic change in how localities addressed problems" (Baum, 1999, p. 289). Therefore, at the least it was expected that while EZ management organizations might well oversee funding, their more important work was to "intervene in policy, changing how public agencies, private firms, nonprofit organizations, and community groups interacted. They would develop relationships among diverse groups and organizations with interests in similar problems, so they could begin to think of themselves as a common domain, define common problems, and set common directions" (Baum, 1999, p. 289). As a new form of community development, it was hoped that Empowerment Zones would create the kind of networks of relationships that Putnam (1995) described as being critical to the creation of social capital.

Putnam's New Turn: Creating Networks of Civic Engagement

As a political scientist, Putnam (1993a) is interested in how social capital functions in the context of democratic institutions. Putnam's twenty-year study of government restructuring in Italy led him to conclude that the differential government performance and overall economic success evident in different regions could be explained by social capital. Putnam concluded that the "norms and networks of civic engagement contribute to economic prosperity and are in turn reinforced by that prosperity" (1993a). Norms of reciprocity are key because they reflect the willingness of people to help one another with the expectation that they can, in turn, call for help. Putnam uses the concept of social capital to describe the reserves of mutual assistance created through norms and networks of civic engagement. In his analysis, reserves of mutual assistance that are accumulated from one set of activities can be applied to other activities, and in the process, mobilized further collective action.

Putnam's interpretation has provoked a rethinking of community economic development. It has challenged conventional analyses that explain and predict success in economic competitiveness by factors such as the development of human capital through workforce training, and other investments (Wallis, Crocker & Schechter, 1998). Putnam's (1993a) analysis suggests that social capital must be seen as a basic element in building a civil society required for economic prosperity. This is because "economics does not predict civics, but civics does predict economics, better than economics itself" (p. 157). Putnam suggests that while social capital is key the development of civil society, it is slow to accumulate and rapidly destroyed. The challenge facing modern societies is to identify and foster the development of new types of civic associations capable of generating new reserves of social capital. Such associations might develop what Putnam refers to as 'bridging capital,' which connects with a broader civic agenda the local social capital, accumulated in the course of informal social interactions that people living in communities engage in. 'Bridging' in this conception implies that social capital serves a catalytic role in mobilizing other forms of capital: financial, physical, and human, towards attaining larger social objectives.

Putnam's connection of social capital with a civic ethic has attracted the critical and supportive attention of social scientists and community activists involved in both economic and civic development. Some regional planners agree with Wilson's (1997) claim that the idea that social capital creates local economic prosperity has lent legitimacy to "what those involved in community economic development have know intuitively for years: the level of inter-personal trust, civic engagement and organizational capacity in a community counts" (p. 745). For those concerned with community economic development the "social capital literature gives the issue [of civic engagement] a more compelling rationale for urgency: the bottom line" (Wilson, 1997, p. 745). At the same time, the concept of social capital raises unsettling questions about how it can be created. It is unsettling because building community or social capital is not a technical problem to be solved by experts. Nor is it entirely a problem of resources, because social capital "unlike physical capital (machinery and equipment), financial capital and human capital, is free – it requires no natural resources, no machines, no bricks and mortar, no paid labor" (Wilson, 1997, p. 746). Community development theorists agree that it does require the development of the kind of trust and openness among community members that Dewey (1916b) described as central to developing democratic society. Putnam (1993a) describes this public good capacity of social capital in the following way:

> Stocks of social capital, such as trust, norms, and networks, tend to be self-reinforcing and cumulative. Virtuous circles result in social equilibrium with high levels of cooperation, trust, reciprocity, civic engagements, and collective well being. These traits define the civic community (p. 177).

Economic development theorists agree with Putnam's (1993b) analysis that productive social capital increases a community's productive potential by promoting business networking. It fosters "business networking; shared leads, equipment and services; joint ventures, faster information flows and more agile transactions" (Wilson, 1997, p. 746). Some distinguish productive social capital as generating compassion and an inclusive sense of community (Etzioni, 1994). Putnam (1993b) acknowledges that unproductive social capital, built on a group's efforts to protect its interest, creates fear and mistrust.

DISCUSSION: ADDRESSING CHALLENGES OF
SOCIAL AND ECONOMIC INTEGRATION AT
THE SCHOOL/COMMUNITY LEVEL

Identifying New Frameworks of Civic Engagement

For community development theorists the challenge has been to identify frameworks of association that can foster civic engagement and a collective sense of responsibility. These frameworks foster the development of social capital as a public good. As described by Putnam (1993a) productive social capital is generated when denser networks of civic engagement are created "that facilitate communication and improve the flow of information about trustworthiness of individuals" (p. 174). Putnam observes that "the denser such networks in a community, the more likely that its citizens will be able to cooperate for mutual benefit" (p. 174). The "dilemma of the commons" which is confronted in efforts to reconcile individual interest with common good, can be overcome more easily in conditions where stocks of social capital, such trust, norms and networks facilitate cooperation and collaboration.

It is exactly these kinds of networks of relationships that are targeted by the public idea that communities across the United States must develop comprehensive school-to-work transition systems in order to meet the agenda of creating a high skilled workforce prepared to work in an economy characterized by rapid globalization and technological change.[27] Crowson's (1992) comments, quoted at the beginning of this chapter suggest than a decade ago the idea that schools are key to community development was only beginning to enter the dialogue in educational scholarship, and only tangentially evident in those disciplines of scholarship concerned with the economic and social contexts of American communities. I have shown in this chapter that key elements of Dewey's conception of education through the occupations can be found in the goals targeted by the STWOA. I argued that Dewey's framework leads us to consider this to be a project of developing civil society in communities by fostering networks of relations that create social capital. Dewey's framework, therefore, provided the link for my turn to examine Putnam's (1993a, b, 1995, 1996) conceptions of social capital as a vehicle for building civil society. The arguments of both Dewey and Putnam, thus, support my conclusion that scholars and policymakers need to view the potential for schools to contribute to social/ economic integration as a challenge in community development.

These are not new claims. Indeed in his review of theoretical approaches for a global community, Blakely (1989) argues that the central concept of community has always inferred psychological, economic, and social relationships. Moreover, communities within the same geographic region have built integrated economic systems. The twin concepts of social and economic solidarity provided a foundation for theory building based on the notion of area economic growth, which sought to determine the factors and direction of development (growth pole), or integrative (functional or decentralized integration) theories that linked communities in the same areas to a common destiny (p. 312).

These theories of community place social mobilization at the core of development. The concept of community embodied in a set of institutions, peopled with shared aspirations and identity was fundamental to community development. Putnam (1993a) and others have revealed that the civic engagement reflected in this notion of community is weak. The problem of development is complicated by the fact that traditional human interacting communities cannot be recreated. Surrogate communities that must be fostered will likely lack the traditionally recognized elements of community such as shared space, common heritage, or intergroup relations. Communities are now formed as networks seeking economic, political and other relationships far beyond geographical boundaries of a particular community. In fact, the new context of globalization means that traditional conceptions of community can no longer serve as the commonly understood base for conceiving of development strategies. The new theories of community development must, therefore, consider the absence of geographic or territorial consciousness of community.

These theories of development must build social capital through fostering the creation of whole new networks of institutions that serve individuals in their various roles. According to Blakely (1986), the new context of development is characterized by institutional adhocracy. In this context community is re-conceptualized as "aspatial solidarity units" and community development becomes a process of network creation and institutional capacity building to fulfill both collective and individual needs. It involves forming networks of networks, and mobilizing those networks by appealing to a sense of common geographic destiny (Blakely, 1986, p. 329). Blakely observes that "aspatial community development techniques require a re-conceptualization of the goals and processes of interaction from strategies designed to build individual connections to strategies designed to build institutional networks that support individual action" (p. 329). He argues that as localities become part of the "space of flows, contemporary theory must deal simultaneously with behavior and groups, [as well as] the relationships among groups within the socioeconomic system" (p. 329). These new conditions require new approaches to community development that focus on the development of temporary institutions responsive to exiting social/economic needs. Building community networks should be based on principles of "aspatial information exchanges" that reflect that fact that networks are formed of concerns and interests of individuals not from geographical elements, and that these interests cross institutional boundaries.

New directions in theories of community development offer guidance to schools involved in developing STW opportunity systems. These theories demand that we reframe how we think about communities and their development in ways that ensure that the partnerships promoted by the STWOA are truly effective in building the social capital that revitalizes communities and ensures their economic prosperity. At the heart of these theories is a new theory of local action. Localism remains key to community development, however, it is a localism that reflects "a quality of action rather than the locus of action" (Blakely, 1986, p. 332).

The criteria guiding local actions must emerge from the development of a shared vision of an outcome that builds civil society. Community development is a value-forming activity. The values that undergird the goals of the School-to-Work Opportunities reflect the kind of democratic education through the occupations that Dewey (1916b) described. Those are values in which communities take responsibility for the development of the education and career opportunities of all youth. Value forming processes of development build networks of relationships that in turn foster the kind of civic engagement that focuses on the common good of the entire community. Schools, involved in this new developmental role do not bowl alone, nor do they or their business and labor partners allow the youth of their communities to search alone for pathways to successful transition from school to work. Rather all participate in the bowling league of the new American economy, a league that can build networks of engagement that will foster a new spirit civic engagement and create new forms of civil society.

NOTES

1. Crowson was reacting to Seymour Sarason's warning that because schools are intractable that failure of educational reform is predictable. See Seymour B. Sarason (1990). *The predictable failure of educational reform.* San Francisco: Jossey-Bass Publishers, pp. 1–8.

2. Lynn Olson first labeled the range of ideas that by the late 1990s had coalesced into a consensus that spurred the passage of the School-To-Work Opportunities Act (1994). See L Olson (1997) *The school- to-work revolution.* Reading, MA: Addison-Wesley.

3. 1998 Report to Congress on Implementation of the School-to-Work Opportunities Act. (http://www.stw.ed.gov/Database/Subject2.cfm?RECNO=2843)

4. For an in-depth analysis of Driscoll and Kerchner's (1999) arguments see Mawhinney, (in press).

5. Similar arguments are presented and applied to a case study in Mawhinney & Kerchner (1997).

6. Mawhinney (in press), drawing on the work of social ecologists, argues that such theorizing must be placed in the context of the broader conceptual framework that focuses attention on the ecology of social relations. The key to social capital formulation as conceptualized by Coleman and extended by Driscoll and Kerchner lies in its attention to active relationships rather than structures. Mawhinney argues that this attention demands an explicit recognition of the microecology of social capital formation.

7. See Mawhinney (1993) for an application of the Advocacy Coalition Framework to two Canadian educational policy changes.

8. The notion that public ideas represent an analytic construct which, according to Mark Moore "function as a broad mandate authorizing and guiding actions by many different organizations" (Moore, 1988, p. 75).

9. While the STWOA received bipartisan support, it has also provoked opposition, in particular, Lynn Olson (1997) reported that the Eagle Forum, a conservative group, views the program "as a conspiracy by big business and big government to shape children's futures charging that it will further vocationalize education and dilute academics" (p. 26). Olson also claims that some teachers "question corporate motives and wonder how much we want businesses involved in the schools" (p. 26).

10. For a description of the background to the passage of the STWOA (1994) see chapters by Robert Reich, Senator Paul Simon, Senator Nancy Langdon Kassebaum, and Hilary Pennington in Jack Jennings (1995) *National issues in education: Goals 2000* and School-to-Work: Bloomington, IN: PDK International (1995). Also see Lynn Olson's (1995) book, *The school-to-work revolution.* Reading, MA: Addison-Wesley.

11. At least one evaluation report states that the diversity of approaches being developed complicates efforts at comparative analysis (see Erilichson & Van Horn, 1999). Other evaluation reports address this challenge by reporting on exemplary STW models, practices, and strategies through case studies of initiatives under way (see School-to-Work Outreach Project Exemplary Model/Practice/Strategy Profiles: (http://www.ici.coled.umn.edu/schooltowork/profiles.html).

12. 1998 Report to Congress on Implementation of the School-to-Work Opportunities Act. (http://www.stw.ed.gov/Database/Subject2.cfm?RECNO=2843)

13. 1998 Ibid

14. See Powers (2000)

15. Ibid

16. Ibid

17. Among others see Wirth (1991) for a description of the nature of the issues in the vocational-liberal studies controversies that sparked debates between John Dewey and the social efficiency philosophers like David Snedden and Charles Posser during the period of the active discussions around the proposals of the vocational education movement (1900–1917).

18. See also, Olson's (1997, p. 26) description of opposition to STW by teachers.

19. See Jack Jennings (1995) *National issues in education: Goals 2000 and School-to-Work*, Bloomington, IN: PDK International. Also see Lynn Olson's (1995) book, *The school to work revolution.* Reading, MA: Addison-Wesley.

20. W. Norton Grubb (1995) makes explicit the influence that he has taken from Dewey in his two volume examination titled from Dewey's call for education *through* the occupations.

21. See research that has been funded through support from STWOA: *1998 National Survey of Local School-to-Work Partnerships* authored by Hulsey, Van Noy & Silverberg (1999) and the *School to Work Governance: A National Review*, authored by Erlichson & Van Horn (1999), and the only cost benefit analysis conducted on the STWOA to date: the *Nebraska School-to Career: An Evaluation*, by Bernier, Lopez, Miller and Partin (1998). See also research reported in academic journals such as Bailey, Hughes and Barr, 2000; Cappelli, Shapiro and Shumanis, 1998, Stasz and Brewer, 1998, Wentling and Waight, 1999.

22. See particularly Driscoll and Kerchner's (1999) analytic essay and Mawhinney's (in press) conceptual framing of the microecology of social capital formation in application to a case study of a magnet high school focusing on school-to-work transitions.

23. School-to-Work Office: http:www.stw.ed.gov.research/goodfacts.htm

24. School-to-Work Office: http//www.stw.ed.gov.research/goodfacts.htm

25. For a description of the elements of Comprehensive Community Initiatives see Vidal (1992).

26. See Gittell & Newman (1998) for a description of the key elements of community capacity in the implementation of Empowerment Zone initiatives.

27. The report: *America's Choice: High Skills or Low Wages* (Commission on the Skills of the American Workforce, 1990), articulates the core elements underlying this public idea.

REFERENCES

Bailey, T., Hughes, K., & Barr, T. (2000). Achieving scale and quality in School-to-Work internships: Findings from two employer surveys. *Educational Evaluation and Policy Analysis*, *22*(1), 41–64.

Baum, H. S. (1999). Education and the empowerment zone: Ad hoc development of an interorganizational domain. *Journal of Urban Affairs*, *21*(3), 289–308.

Beck, L. G., & Murphy, J. (1996). *The four imperatives of a successful school*. Thousand Oaks, CA: Corwin.

Bernier, R., Lopez, G., Miller, J., & Partin, S. (1998). Nebraska School-to-Career: An Evaluation Lincoln, NE: Department of Economic Development (http://www.stw.ed.gov/Database/Subject2.cfm?RECNO=1247).

Blakely, E. J. (1989). Theoretical approaches for a global community. In: J. A. Christenson & J. W. Robinson, Jr. (Eds), *Community Development in Perspective* (pp. 307–336), Ames, IO: Iowa State University Press.

Bourdieu, P. (1977). Cultural reproduction and social reproduction. In: J. Karabel & A. H. Halsey (Eds), *Power and Ideology in Education*. New York: Oxford University Press.

Bourdieu, P., & Passeron, J. C. (1977). *Reproduction in education, society and culture*. Beverly Hills, CA: Sage.

Bourdieu, P. (1986). The forms of capital. In: J. G. Richardson (Ed.), *Handbook of Theory and Research for the Sociology of Education* (pp. 241–258). New York: Greenwood Press.

Boyd, W., & Crowson, R. L. (1996). Structures and strategies: Toward and understanding of alternative models for coordinated children's services. In: J. C. Cibulka & W. Kritek (Eds), *Coordination Among Schools, Families, and Communities: Prospects for Reform* (pp. 137– 169). Albany: SUNY Press.

Boyd, W., Crowson, R. L., & Gresson, A. (1997). Neighborhood initiatives, community agencies, and the public schools: A changing scene for development and learning of children. In: M. C. Wang & M. C. Reynolds (Eds), *Development and Learning of Children and Youth in Urban America* (pp. 81–103). Philadelphia: Temple University Center for Research in Human Development and Education.

Braun, E., & Busch, A. (Eds) (1999). *Public Policy and Political Ideas*. Cheltenham, UK: Edward Elgar.

Cappelli, P., Shapiro, D., & Shumanis, N. (1998). Employer participation in school-to-work programs. *Annals, AAPSS, 559*(September), 109–124.

Coleman, J. (1990). *Foundations of social theory*. Cambridge, MA & London, England: The Belknap Press of Harvard University Press.

Commission on the Skills of the American Workforce (1990). *America's choice: High skills or low wages*. (Washington, D.C.: National Center on Education and the Economy.

Crowson, R. L. (1992). *School-community relations under reform*. Berkeley, CA: McCutchan.

Crowson, R. L., & Boyd, W. L (1993). Coordinated services for children: Designing arks for storms and seas unknown. *American Journal of Education, 19*(2), 140–178.

Crowson, R. L., Wong, K. K., & Aypay, A. (2000). The quiet reform in American education: Policy issues and conceptual challenges in the school-to-work transition. *Educational Policy, 14* (2), 241–258.

Dewey, J. (1915). Education vs. trade training: Dr. Dewey's reply. *The New Republic, 3*, 15 May.

Dewey, J. (1916a). Comment. *The New Republic, 3*, 5 May.

Dewey, J. (1916b). *Democracy and education: An introduction to the philosophy of education*. New York: Macmillan.

Dewey, J. (1950). *Reconstruction in philosophy*. The American Library: New York. (1960 edition).

Driscoll, M. R., & Kerchner, C. T. (1999). The implications of social capital for schools, communities, and cities: Educational Administration as if a sense of place mattered. In: J. Murphy & K. S. Louis (Eds), *Handbook of Research on Educational Administration* (pp. 385–404). San Francisco, CA: Jossey-Bass Publishers.

Edelman, I. (2000). The riddle of community-based initiatives. *National Civic Review, 89*(2), 169–181.

Erlichson, B. A., & Van Horn, C. E. (1999). School to Work Governance: A National Review. Philadelphia, PN: Consortium for Policy Research in Education. (http://www.stw.ed.gov/ Database/Subject2.cfm?RECNO=1269).

Etzioni, A. (1994). *Spirit of community: The reinvention of American society*. New York: Simon and Schuster.

Foley, M., & Edwards, B. (1997). Escape from politics? Social theory and the social capital debate. *American Behavioral Scientist, 40*(5), 553.

Gittell, M., & Newmann, K. (1998). *Empowerment Zone implementation: Community participation and community capacity*. New York: City University of New York, Graduate School and University Center.

Grubb, W. N. (1995). Resolving the paradox of the high school. In: W. N. Grubb (Ed.) *Education Through the Occupations in American High Schools*, Vol. 1: *Approaches to Integrating Academic and Vocational Education*. (pp. 1–8). New York: Teachers College Press.

Halpern, R. (1995). *Rebuilding the Inner City*. New York: Columbia University Press.

Henry, M. (1996). *Parent-school collaboration: Feminist organizational structures and school leadership*. Albany, NY: State University of New York Press.

Hulsey, L, Van Noy, M., & Silverberg, M. (1999). *1998 National Survey of Local School-to-Work Partnerships*. Princeton, NJ: Mathematica Policy Research, Inc. (http://www.stw.ed.gov/ Database/Subject2.cfm?RECNO=1134).

Kincheloe, J. L. (1999). *How do we tell the workers? The socioeconomic foundations of work and vocational education.* Boulder, CA: Westview Press.

Knapp, M. S. (1995). How shall we study comprehensive, collaborative services for children and families? *Educational Researcher, 24*(4), 5–16.

Lemann, N. (1996, April). Kicking in groups. *Atlantic Monthly,* p. 25.

Malen, B., Ogawa R. T., & Kranz, J. (1989). What do we know about school-based management? A case study of the literature and a call for research. Paper presented at the conference on choice and control in American Education. In: W. Clune & J. Witte (Eds), *Choice and Control in American Education, Vol. 2: The Practice of Choice, Decentralization and School Restructuring* (pp. 289–342). New York: Falmer.

Martinez- Brawley, E. (1995). Community. *Encyclopedia of Social Work.* Washington, DC: NASW Press.

Mawhinney, H. B. (1993). An advocacy coalition approach to policy change in Canadian education. In: H. Jenkins-Smith & P. A. Sabatier (Eds), *An Advocacy Coalition Model of Policy Change and Learning* (pp. 91–128). Boulder, CO: Westview Press.

Mawhinney, H. B. (1996). Institutional effects of strategic efforts at community enrichment. In: J. G. Cibulka & W. J. Kritek (1996) *Coordination Among Schools, Families, and Communities: Prospects for Educational Reform* (pp. 223–243) Albany, NY: SUNY Press.

Mawhinney, H. B. (1998). School wars or school transformation: Professionalizing teaching and involving communities. *Peabody Journal of Education, 73,* 36–55.

Mawhinney, H. B. (in press). The Microecology of social capital formation: Developing community beyond the schoolhouse door. In: G. Furman-Brown (Ed.), *School as Community: From Promise to Practice.* Albany: SUNY Press.

Mawhinney, H. B., & Kerchner, C. (1997). The micro-ecology of school community links. In: M. McClure & J. C. Lindle (Eds), *Expertise Versus Responsiveness in Children's Worlds: Politics in School, Home and Community Relationships* (pp. 29–36). Washington, D.C.: Falmer Press.

Mawhinney, H. B., & Smrekar, C. (1996). Institutional constraints to advocacy in school-community collaborations. *Educational Policy, 10,* 480–501.

Merz, C., & Furman, G. (1997). *Community and schools: Promise and paradox.* New York: Teachers College Press.

Moore, M. (1988). What sort of ideas become public ideas? In: R. Reich (Ed.), *The Power of Public Ideas* (pp. 70–83) Cambridge, MA: Ballinger.

Murphy, J. (1999). New consumerism: Evolving market dynamics in the institutional dimension of schooling. In: J. Murphy & K. S. Louis (Eds), *Handbook of Research on Educational Administration* (pp. 405–419). San Francisco, CA: Jossey-Bass Publishers.

National Center on the Educational Quality of the Workforce. Advisory Board (1995). *Making good jobs for young people a national priority.* Philadelphia: University of Pennsylvania, National Center on the Educational Quality of the Workforce.

National Commission on Excellence in Education (1983). *A nation at risk: The imperative for educational reform.* Washington, DC: U.S. Government Printing Office.

National School-to-Work Office (1998) *The National School-to-Work Office's Annual Report to Congress.* Washington, DC: (http://www.stw.ed.gov/Database/Subject2.cfm?RECNO=2843)

Olson, L. (1997). *The school-to-work revolution: How employers and educators are joining forces to prepare tomorrow's skilled workforce.* Reading, MA: Addison-Wesley.

Plant, R. (1991). *Modern political thought.* Oxford, U.K. Basil Blackwood.

Powers, S. J. (2000, October 11). *School-to-Work 2000: Thinking About Tomorrow Conference.* Washington, DC: National School-to-Work Office (http://www.stw.ed.gov/conf/sp_remarks.htm)

Prosser, C., & Quigley, T. (1950). *Vocational education in democracy.* Chicago: American Technical Society.

Putnam, R. D. (1993a). *Making democracy work: Civic traditions in modern Italy.* Princeton, NJ: Princeton University Press.

Putnam, R. D. (1993b). The prosperous community: Social capital and economic growth. *American Prospect,* (Spring), 35–42.

Putnam, R. D. (1995). Bowling alone: America's declining social capital. *Journal of Democracy,* (January), 65–78.

Putnam, R. D. (1996). The strange disappearance of civic America. *American Prospect, 24*(Winter), 34–48.

Reich, R. (Ed.) (1988). *The Power of Public Ideas.* Cambridge: Harvard University Press.

Reich, R. B. (1995). Building a framework for a school-to-work opportunities system. In: J. F. Jennings (Ed.), *National Issues in Education: Goals 2000 and School-to-Work* (pp. 119–125). Washington, DC: Phi Delta Kappa International.

Sabatier, P. A., & Jenkins-Smith, H. C. (Eds) (1993). *Policy Change and Learning: An Advocacy Coalition Approach.* Boulder, CO: Westview Press.

Sabatier, P. A. (Ed.) (1999). *Theories of the Policy Process.* Boulder, CO: Westview Press.

Saul, R. S. (1998). On connecting school and work. *Annals, AAPSS, 559,* 168–175.

Schorr, L. (1988). *Within our reach: Breaking the cycle of disadvantage.* New York: Anchor.

Schorr, L. (1997). *Common purpose: Strengthening families and neighborhood to rebuild America.* New York: Anchor.

Secretary's Commission on Achieving Necessary Skills. (1991). *What work requires: A SCANS report for America 2000.* Washington, D.C.: U.S. Department of Labor.

Shanker, A. (1995, 23 April). *Linking school and work.* New York Times.

Smrekar, C. (1996). *The impact of school choice and community.* Albany: State University of New York Press.

Smrekar, C. E., & Mawhinney, H. B. (1999). Integrated services: Challenges in linking schools, families, and communities. In: J. Murphy & K. S. Louis (Eds), *Handbook of Research on Educational Administration* (pp. 443–461). San Francisco, CA: Jossey-Bass Publishers.

Stasz, C., & Brewer, D. J. (1998). Work-based learning: Student perspectives on quality and links to school. *Educational Evaluation and Policy Analysis, 19*(1), 31–46.

Stengel, R. (1996, July 22). Bowling together. *Time,* p. 35.

Stone, D. A. (1988). *Policy paradox and political reason.* New York: Harper Collins.

The School-to-Work Opportunities Act (1994). U.S. Code Service, Public Law 103–239, HR 2884.

Thurow, L. (1985). *The zero-sum solution.* New York: NY: Simon & Schuster.

Vidal, A. (1992). *Rebuilding communities: A national study of Community Development Corporations.* New York: New School for Social Research.

Wallis, A., Crocker, J. P., & Schechter, B. (1998). Social capital and community building: Part one. *National Civic Review, 87*(3), 253–271.

Warren, R. L. (1975) Comprehensive planning and coordination: Some functional aspects. *Social Problems, 20,* 335–364.

Wentling, R. M., & Waight, C. S. (1999). Barriers that hinder the successful transiton of minority youth into the workplace. *Journal of Vocational Education Research, 24*(4), 165–183.

William T. Grant Foundation Commission on Work, Family, and Citizenship. (1988). *The forgotten half: Pathways to success for America's youth and young families*. Washington, DC: William T. Grant Foundation.

Wilson, P. A. (1997). Building social capital: A learning agenda for the twenty-first century. *Urban Studies, 34*(5–6), 745–760.

Wirth, A. G. (1991). Issues in the vocational-liberal studies controversy (1900–1917): John Dewey vs the social efficiency philosophers. In: D. Corson (Ed.), *Education for Work: Background to Policy and Curriculum* (pp. 55–64). Clevedon: Multilingual Matters Ltd.

Zemsky, R., Shapiro, D., Iannozzi, M., Cappelli, P., & Bailey, T. (1998). *The transition from initial education to working life in the United States of America*. Stanford, CA: National Center for Postsecondary Improvement.